BIOTRANSFORMATION AND FATE OF CHEMICALS IN THE AQUATIC ENVIRONMENT

BIOTRANSFORMATION AND FATE OF CHEMICALS IN THE AQUATIC ENVIRONMENT

Proceedings of a Workshop Held at the
University of Michigan Biological Station
Pellston, Michigan
14-18 August 1979

EDITORS

Alan W. Maki
Procter & Gamble Company

Kenneth L. Dickson
North Texas State University

John Cairns, Jr.
Virginia Polytechnic Institute
and State University

American Society
for Microbiology

Washington, DC 20006

Library of Congress Cataloging in Publication Data

Main entry under title:
Biotransformation and fate of chemicals in the aquatic
  environment.

  Proceedings of a workshop held Aug. 14–18, 1979 at
the University of Michigan Biological Station, Pellston,
Mich.
  Includes index.
  1. Biodegradation—Congresses. 2. Aquatic micro-
biology—Congresses. 3. Aquatic ecology—Congresses.
4. Water—Pollution—Measurement—Congresses. I. Maki,
Alan W., 1947-   II. Dickson, Kenneth L.
III. Cairns, John, 1923-
QH530.5.B56   574.5'263   80-16522

ISBN 0-914826-28-X

CONTENTS

Acknowledgments

The financial support for conduct of the workshop and publication of these proceedings was provided through a grant-in-aid from the Chemical Manufacturers Association to North Texas State University and is gratefully appreciated.

The co-chairmen thank each of the participants for their commitment to the success of the workshop. We are also indebted to the staff, particularly Mark Paddock and Jim Daunter, of the University of Michigan Biological Station and D. M. Gates, Director, for coordinating the arrangements and accommodations for the Workshop. We express our gratitude to Teresa M. Daunter, Jean O. Cairns, and Heather Cairns for typing support and preparation of the early drafts of discussion papers. Finally, we express our sincere appreciation to Terra Clem for typing and collating these proceedings.

Workshop Participants

George Baughman
Southeast Environmental Research Laboratory, USEPA, Athens, Georgia 30605

Harold Bergman
Department of Zoology, University of Wyoming, Laramie, Wyoming 82071

Gary Blau
DOW Chemical Environmental Science Research, Midland, Michigan 48640

Dean Branson
Dow Chemical Company, Midland, Michigan 48640

Bob Brink
USEPA Office of Toxic Substances, Washington, D.C. 20406

John Cairns, Jr.
Center for Environmental Studies, Virginia Polytechnic Institute and State University, Blacksburg, Virginia 24061

Lenore S. Clesceri
Department of Biology, Rensselaer Polytechnic Institute, Troy, New York 12181

Ken Dickson
Institute of Applied Sciences, North Texas State University, Denton, Texas 76203

David T. Gibson
University of Texas at Austin, Austin, Texas 78712

William E. Gledhill
Monsanto Industrial Chemicals Company, St. Louis, Missouri 63166

Jerry L. Hamelink
Lilly Research Laboratory, Greenfield, Indiana 46140

B. Thomas Johnson
Columbia National Fishery Research Laboratory, Columbia, Missouri 65201

Howard E. Johnson
Department of Fisheries and Wildlife, Michigan State University, East Lansing, Michigan 48823

Masaru Kitano
Kaga Kuhin Kenza Kyokai, Kaga Kuhin Anzen Center, 4-1-1 Higashi-Mukohjima, Sumida-Ku, Tokyo, 131, Japan

Ernest Ladd
FMC Corporation, Philadelphia, Pennsylvania 19103

Bob Larson
Environmental Safety Department, The Procter & Gamble Company, Cincinnati, Ohio 45217

Chris Lee
Unilever Research Laboratory, Port Sunlight, Wirral, Merseyside L62 4XN, England

G. Fred Lee
Department of Civil Engineering, Colorado State University, Fort Collins, Colorado 80523

Gordon Loewengart
Allied Chemical Company, Morristown, New Jersey 07960

A. W. Maki
Environmental Safety Department, The Procter & Gamble Company, Cincinnati, Ohio 45217

Fumio Matsumura
Pesticide Research Center, Michigan State University, East Lansing, Michigan 48823

Doris F. Paris
Southeast Environmental Research Laboratory, USEPA, Athens, Georgia 30605

Hap Pritchard
USEPA–ERL, Sabine Island, Gulf Breeze, Florida 32561

J. Spain
USEPA–ERL, Sabine Island, Gulf Breeze, Florida 32561

Arthur Stern
USEPA Office of Toxic Substances, Washington, D.C. 20460

James M. Tiedje
Department of Microbiology and Public Health, Michigan State University, East Lansing, Michigan 48823

V. Zitko
Fisheries and Environmental Sciences, Biological Station, Saint Andrews, New Brunswick E0G 2X0, Canada

Chapter 1. INTRODUCTION

KENNETH I. DICKSON, ALAN W. MAKI, AND JOHN CAIRNS, JR.

To fully understand the contents of this book and the workshop from which it emanates, a review of activities which influenced the contents is helpful. The Toxic Substances Control Act (TSCA) was passed by Congress in October 1976, stipulating that no person may manufacture or process a chemical substance for a new use without obtaining clearance from the U.S. Environmental Protection Agency. TSCA represents an attempt to establish a mechanism whereby the hazard of a chemical substance to human health and the environment can be assessed before the chemical is introduced into the environment. TSCA requires premarketing testing and a judgment by the Environmental Protection Agency and industry on the degree of risk associated with the use of a chemical substance. In response to this stimulus, the scientific community in industry, academia, and government began a search for a conceptual framework around which to organize the hazard evaluation process. Hazard evaluation approaches were needed by industry as part of its product evaluation process and by regulatory agencies charged with the final determination of the risk associated with the manufacture, use, and distribution of a chemical substance.

As the development of test methods proceeds and as new environmental regulations are written, the commonality of data needs and inputs between these regulations has become increasingly clear. Although written for distinct and individually important purposes, TSCA, the Clean Water Act and amendments, the Water Quality Criteria, the Resource Conservation and Recovery Act (RCRA), the Federal Insecticide, Fungicide and Rodenticide Act (FIFRA), and numerous other regulations all have many common data needs considering toxicology and exposure assessment. The conveners of the workshop and the editors of this book expect that efforts such as this workshop should help to develop both the commonality of data needs and a conceptual framework for chemical and waste evaluations that will eventually serve as the basis for much of the currently evolving environmental safety legislation.

In an effort to assist in the development of this conceptual framework for conducting an evaluation of the hazard of chemicals to aquatic life, an ad hoc workshop was conducted on 13–17 June 1977 at the University of Michigan's Biological Station, Pellston, Mich., entitled "Application of Toxicity Testing Methods as Predictive Tools for Aquatic Safety Evaluation." Twenty-seven invited participants, including representatives from industry, universities, and regulatory agencies, addressed the "state of the art" of aquatic toxicology and environmental safety assessment. In addition to identifying the methods and tools of aquatic toxicology appropriate for determining the effects of chemicals on aquatic life, the participants helped to formulate the concept that the hazard of a chemical can be determined by relating the predicted or estimated environmental exposure concentration to the measured concentration which causes an adverse effect as determined through the conduct of laboratory toxicity tests. To predict environmental exposure concentrations of a chemical, knowledge of the physical, chemical, and biological processes which control its fate is essential. The proceedings from this workshop were published by the American Society for Testing and Materials in the book *Estimating the Hazard of Chemical Substances to Aquatic Life* (1).

The participants of the 1977 Pellston workshop recognized that since the process of safety evaluation of new chemicals was truly a global problem, it would be beneficial to hold a subsequent workshop which would examine additional hazard assessment approaches from this country, as well as approaches being developed in Europe and Japan. Such a workshop was held on 13–18 August 1978 at Waterville Valley, N.H. Like its predecessor, this workshop included representatives from industry, universities, and governmental agencies, who interacted to analyze and revise the state of the art in assessing the hazards of chemicals to aquatic life. The results of this workshop, *Analyzing the Hazard Evaluation Process*, published by the American Fisheries Society (2), identified the need for continued effort to develop methods to assess the fate of chemicals in the aquatic environment. Since an estimate of the environmental exposure concentration is essential to a meaningful hazard evaluation, reliable methods are required for those chemical, physical, and biological processes which determine fate. Processes identified by the 1978 Waterville Valley workshop as important in determining the fate of a chemical included photochemical degradation, chemical degradation, evaporation, sediment binding, uptake and depuration by organisms, and microbiological transformation.

The 1977 Pellston workshop and the 1978 Waterville Valley workshop were extremely useful in addressing state-of-the-art testing methods and hazard evaluation schemes. Both indicated that considerable effort was needed in the development of methods to predict the environmental fate of chemicals. For this reason, a third workshop was planned, organized, and conducted around the theme "Biotransformation and Fate of Chemicals in the Aquatic Environment." This workshop was held on 14–18 August 1979 at the University of Michigan's Biological Station in Pellston; it focused on the methodology for biodegradation testing of chemicals and considered how predictive modeling contributes to fate determinations. The proceedings of this workshop constitute the contents of this book.

BACKGROUND

This book had its origin in the ideas and questions of the participants in the 1977 Pellston and 1978 Waterville Valley workshops. To plan for the 1979 Pellston workshop, two ad hoc planning meetings were held to identify specific program topics and to recommend a list of participants. The first planning committee was assembled in Chicago on 13 October 1978. The members and affiliations of the planning committee were as follows:

Art Stern
Environmental Protection Agency
Washington, D.C.

Bill Gledhill
Monsanto
St. Louis, Mo.

Hap Pritchard
Environmental Protection Agency
Gulf Breeze, Fla.

Jim Tiedje
Michigan State University
East Lansing, Mich.

Peter Chapman
University of Minnesota
Minneapolis, Minn.

Bob Larson
Procter & Gamble
Cincinnati, Ohio

Al Maki
Procter & Gamble
Cincinnati, Ohio

The committee members were selected on the basis of their backgrounds and expertise in the field of biodegradation testing and biotransformation of

To integrate the topics identified by this planning committee into an overall program for determining the environmental fate of chemicals, an additional committee was assembled. This group met in New Orleans on 18 October 1978 and consisted of the following:

George Baughman
Environmental Protection Agency
Athens, Ga.

Ken Macek
EG&G Bionomics
Wareham, Mass.

Dean Branson
Dow Chemical Co.
Midland, Mich.

Ken Dickson
North Texas State University
Denton, Tex.

Alan Maki
Procter & Gamble
Cincinnati, Ohio

Suggestions from both planning groups were incorporated into the final workshop design and program by the following ad hoc committee, which met on 9 January 1979 in Atlanta, Ga.:

George Baughman
Environmental Protection Agency
Athens, Ga.

John Cairns, Jr.
Virginia Polytechnic Institute and State
   University
Blacksburg, Va.

Steve Herbes
Oak Ridge National Laboratory
Oak Ridge, Tenn.

Dean Branson
Dow Chemical Co.
Midland, Mich.

Ken Dickson
North Texas State University
Denton, Tex.

Al Maki
Procter & Gamble
Cincinnati, Ohio

Workshop Objectives and Scope

The 1979 Pellston workshop, "Biotransformation and Fate of Chemicals in the Aquatic Environment," had the following objectives:

1. To consider the effects of microbial transformation processes on the environmental fate of chemicals and to examine methods of predictive modeling for fate determination
2. To examine what types of biotransformation data are needed to predict environmental concentrations of chemicals

3. To evaluate methods currently used to describe kinetics of microbial environmental transformations
4. To consider the effects of chemical/physical variables on biological transformations
5. To examine the rationale for extrapolation of laboratory data to the real world

To address these specific objectives, eight individual workshop sessions were conducted. Each session was initiated with one or more well-defined discussion initiation papers from a specific author or group of authors named by the ad hoc planning committees. These papers were presented during plenary sessions of the workshop followed by a discussion period open to all participants. Discussion was led by a session chairman. It was the responsibility of these chairmen and their committees to summarize each session, identifying the major consensus conclusions of the participants regarding topics in each session. The workshop sessions are given below.

Session I. Introduction

A. W. Maki, K. L. Dickson, and John Cairns, Jr.

Participants were presented with the goals and objectives of the workshop. Background information was presented on the relationship of the current workshop to the 1977 Pellston and 1978 Waterville Valley workshops.

Session II. Environmental Exposure

Concentration of Chemicals: Background
Discussion Initiation Paper: G. Fred Lee and R. Anne Jones
Session Chairman: G. Loewengart
Session Summary Committee: G. Loewengart and G. Fred Lee

The main emphasis of this session was a discussion of why knowledge of environmental concentration is important in the hazard evaluation process; it included a consideration of overall scope and the need to know environmental exposure concentrations of chemicals. Why do we need these estimates, predictions, and measurements, and how are these data used in hazard evaluation programs for chemicals? How are environmental chemistry and biotransformation data used in making predictions or estimates of environmental exposure concentrations?

Session III. Evaluation of Methods

Discussion Initiation Paper: B. Thomas Johnson
Discussion Initiation Paper: P. A. Gilbert and C. M. Lee
Session Chairman: W. E. Gledhill
Session Summary Committee: W. E. Gledhill, B. T. Johnson, M. Kitano, and C. M. Lee

Examples of laboratory simulations of microbial environmental compartments—how they are done—were presented. Discussion included the importance of acclimation and concentration effects, the identification of parameters for which we do not have methods, and an identification of analytical chemistry data needed to support these tests. Evaluation criteria were developed, as well as a critical ranking of existing methodology. The session addressed the questions: (i) What current and evolving techniques apply to biodegradation of insoluble materials, anaerobic systems, solvents? (ii) What is the role of microcosms or other multicompartment model systems in biodegradation studies? (iii) What methods exist for testing biodegradation products and when do breakdown products become important? (iv) How does cometabolism influence determination of biotransformation?

Session IV. Bioavailability and the Materials Balance Approach to Estimating Environmental Concentration

Discussion Initiation Paper: D. R. Branson and G. E. Blau
Discussion Initiation Paper: J. L. Hamelink
Session Chairman: R. Brink
Session Summary Committee: R. Brink, G. Blau, and D. Branson

What chemical/physical parameters effect bioavailability of compounds? How is bioavailability operationally defined and what are the effects of adsorption kinetics? What natural environmental parameters affect how a compound is going to degrade? What is the influence of other carbon sources? Does the concept of bioavailability affect or alter a determination of existing environmental concentration for a test compound?

Session V. Environmental Extrapolation of Biotransformation Data

Discussion Initiation Paper: R. J. Larson
Discussion Initiation Paper: L. S. Clesceri
Session Chairman: P. H. Pritchard
Session Summary Committee: P. H. Pritchard, R. J. Larson, and L. S. Clesceri

A discussion of the ways, means, and problems of extrapolating laboratory results to actual environmental systems was conducted. What role does kinetic modeling play in predicting environmental fate? What models are most appropriate for predicting rates of biodegradation in environmental systems? Are there differences in chemical and biological kinetics, especially at low chemical concentration?

Session VI. Quantitative Expression of Biotransformation

Discussion Initiation Paper: G. L. Baughman, D. E. Paris, and J. Spain
Session Chairman: H. L. Bergman
Session Summary Committee: H. L. Bergman, G. L. Baughman, D. F. Paris, and J. Spain

Can the kinetics of microbial environmental transformation be adequately expressed for hazard assessment? What criteria can be used for measuring and arriving at a critique of these methods? What degree of accuracy is needed? What are the criteria for deciding whether a particular approach is or is not valid? What chemical, physical, and biological environmental variables influence the derivation of these expressions?

Session VII. Fate of Chemicals in the Aquatic Environment—Case Studies

Discussion Initiation Paper: J. M. Tiedje
Discussion Initiation Paper: V. Zitko
Discussion Initiation Paper: F. Matsumura
Session Chairman: H. E. Johnson
Session Summary Committee: H. E. Johnson, V. Zitko, and F. Matsumura

The main emphasis of this session was a discussion of three well-studied materials and how test results were used to assess the overall biodegradability or environmental exposure concentration. What sequence of testing was used to yield increasingly refined estimates of biodegradation potential?

Session VIII. Workshop Summary

Session Chairman: J. L. Hamelink
Workshop Summary Committee: D. L. Gibson, G. F. Lee, G. V. Loewengart, and A. M. Stern

The summary committees from each session presented their conclusions and recommendations. An overall workshop summary, conclusions, and identification of areas of future research needs were prepared by the workshop summary committee.

Workshop Conduct

Prior to arrival at the University of Michigan's Biological Station on 19 August 1979, each of the participants received a notebook containing drafts of the discussion initiation papers and other relevant literature. Plenary sessions were conducted on Monday, Tuesday, and Wednesday. The first three nights and all day on Thursday the individual session committees met and formulated their consensus summaries for presentation during the Friday plenary session. Initial drafts of each session summary were presented during the final session on Friday. An overall summary was presented by the workshop summary committee. Final drafts of each of the summaries were completed via subsequent mailings. However, no substantial changes were permitted after the workshop was finished. Every attempt has been made to ensure that editorial review has been carried out in the spirit of the workshop.

The chapters of this book coincide with the workshop sessions. Each chapter, with the exception of this introduction and the Workshop Summary, contains one or more discussion initiation papers and a short summary report prepared by the session summary committee.

LITERATURE CITED

1. **Cairns, J., K. L. Dickson, and A. W. Maki (ed.).** 1978. Estimating the hazard of chemical substances to aquatic life. ASTM Special Technical Publ. 657. American Society for Testing and Materials, Philadelphia.
2. **Dickson, K. L., A. W. Maki, and J. Cairns (ed.).** 1979. Analyzing the hazard evaluation process. American Fisheries Society, Bethesda, Md.

Chapter 2. ENVIRONMENTAL EXPOSURE CONCENTRATION OF CHEMICALS

Role of Biotransformation in Environmental Hazard Assessment

G. FRED LEE AND R. ANNE JONES

Department of Civil Engineering, Environmental Engineering Program, Colorado State University, Fort Collins, Colorado 80523

In accord with the requirements of the Toxic Substances Control Act, new and expanded-use chemicals must be screened for their potential environmental hazard prior to large-scale manufacture and use. This environmental hazard assessment should have two basic components, environmental toxicology and environmental chemistry-fate. Environmental toxicology considers the concentrations (and durations of exposure to organisms) of a chemical and its transformation products which can adversely affect aquatic and terrestrial organisms. Environmental chemistry-fate considers the transport pathways and ultimate disposition of the chemical in the environment and what transformations of the chemical occur and the rates of these transformations. It also provides an estimate of the expected concentrations of a contaminant in various environmental compartments. A hazard assessment is made by proceeding through a series of testing levels or tiers which develop information on the toxicology of the chemical and its environmental chemistry-fate until a decision can be made that the environmental risk associated with the manufacture and use of the chemical is acceptable or unacceptable. One of the major factors to consider in determining an expected environmental concentration of a chemical is the rate and extent of its biotransformation in the environment. Considerable work needs to be done to relate the results of laboratory tests for biotransformation rates to the chemical's behavior under various environmental conditions. This requires coordinated laboratory and field studies conducted so as to determine the factors influencing biotransformation in environmental systems. A discussion is presented of some of the more important factors to consider in evaluating the biotransformation potential and fate data required for an environmental hazard assessment.

In the past decade the world has faced several chemical crises in which widespread contamination has occurred by a chemical thought to have the potential to cause significant environmental degradation or harm to humans. The classic examples of these kinds of problems have been DDT (dichlorodiphenyltrichloroethane) and several other chlorinated hydrocarbon pesticides, polychlorinated biphenyls (PCBs), mercury, and, in the Chesapeake Bay area, Kepone. As a result of these crises, several countries, including the United States, have banned or restricted the manufacture and use of certain chemicals such as PCBs in an effort to reduce the extent of environmental contamination. Further, several countries have passed legislation which requires prescreening of new and some existing chemicals prior to large-scale manufacture and use. In the United States this legislation is called the Toxic Substances Control Act (TSCA). The primary purpose of TSCA is to try to prevent future chemical crises of the DDT, PCB, and Kepone types.

The basic process involved in prescreening new chemicals is the development of an environmental hazard assessment. Such an assessment is usually built on two fundamental types of information, environmental and human toxicology and environmental chemistry-fate. Environmental toxicology considers the concentrations of each potentially significant form of the contaminant under consideration and its transformation products that are potentially adverse to terrestrial and aquatic life, including humans. Environmental chemistry-fate considers the transport and transformation of the chemical for all modes of input, from the point of entry into the environment to its final disposition. The physical processes of advection-transport and dilution-dispersion must be defined for the terrestrial, atmospheric, and aquatic components of the environment. Environmental chemistry-fate also considers the chemical processes that influence the form(s) (chemical species) of the contaminant and its transformation products in each of the major components of the environment. For example, for the aquatic environment, consideration must be given to the forms and concentrations of each form in true solution (i.e., dissolved), associated with particulate matter (such as erosional materials and organic detritus that are suspended in the water column and deposited in the sediments), and found upon or contained within aquatic organisms. There is also the potential for some highly volatile compounds to be transported to or from gas bubbles in the water column. Few chemicals are completely conservative or nonreactive in the environment, i.e., chemicals whose concentrations in the environment change only as a result of physical processes of dilution and dispersion. Most chemicals undergo a variety of transformations, the majority of which can have a pronounced effect on the environmental concentration of the forms of the contaminant.

Major types of reactions that commonly occur in the aqueous environment are acid-base, precipitation, complexation, oxidation-reduction, sorption including biological uptake, hydrolysis, photolysis, gas transfer, and biochemically mediated reactions-biotransformation. This workshop, devoted to biotransformation and the fate of chemicals in the aquatic environment, is designed to review the current information on the role of biotransformation as a mechanism for influencing the environmental concentrations of a potential contaminant in aquatic systems as it relates to an environmental hazard assessment. Emphasis is given to the current knowledge of the assessment of the potential for biochemically mediated reactions to occur in aquatic systems. Of particular concern is information pertinent to predicting the rate and extent of biologically mediated reactions in various parts of the aquatic environment. This paper provides an overview of the environmental hazard assessment process with particular emphasis on how laboratory-derived biotransformation information is used in this process. Also, a number of suggestions are made concerning additional research that is needed to develop appropriate biotransformation information for environmental hazard assessment of a new or expanded-use chemical.

ENVIRONMENTAL HAZARD ASSESSMENT

As discussed above, environmental (and human) toxicology and environmental chemistry-fate information are the backbone of environmental hazard assessment. The typical approach that is being used or suggested for use is discussed in several papers from the previous Pellston workshop (3); an example of this approach is outlined in Fig. 1. In this figure, Kimerle et al. (5) presented a tiered hazard assessment scheme which appears to be typical of those being used today by industry and those proposed for use by governmental agencies. The tiered nature of this approach is generally regarded as highly important in the wise use of funds for prescreening new chemicals. The alternative to the tiered approach is a fixed list of tests which must be performed on every compound irrespective of its characteristics. This latter mechanical approach to hazard assessment does not take into account the fact that individuals knowledgeable in environmental chemistry can readily predict certain aspects of expected environmental behavior based on the structure of the compound or the results of simple, relatively inexpensive laboratory tests.

The tiered screening process for hazard assessment generally involves the measurement of toxicological and chemical properties in a series of levels of sophistication, with a decision point at the end of each level. The decision choices are: (i) do not manufacture because of excessive expected hazard or restrict manufacture and use to reduce environmental hazard to an acceptable level; (ii) proceed with manufacture—acceptable expected hazard; and (iii) continue testing. The first tier of testing usually consists of relatively unsophisticated, inexpensive tests. The second and subsequent tiers involve tests having greater degrees of sophistication for environmental toxicity and chemical behavior. In proceeding through the tiers of a hazard assessment program it is important to recognize that a fixed rate or extent of biotransformation or other chemical reaction cannot be used to trigger the next tier. The trigger for work in higher tiers must be based on an evaluation of all information, including other chemical, toxicological, and economic considerations. As one proceeds through the tiers, the reliability of the estimates of toxicity and environmental concentration, and hence the precision of the decision, should be significantly improving. This relationship is depicted in Fig. 2. It is important to note that highly precise estimates of aquatic toxicity and environmental concentration are not needed in a hazard assessment. The precision of the measurements in any tier should be geared to the sensitivity of the decision-making process.

Figure 2 shows that the estimated environmental concentration is always less than the "no effect" concentration. There will be situations, however, in which the estimated environmental concentration will be greater than the no-effect level; i.e., there will be some expected environmental degradation associated with the manufacture or use, or both, of a particular chemical. Under these conditions it may be determined that it is in the best overall interest of society to allow some degradation of the environment in order to acquire the chemical's benefits. An example of this type of chemical would be one used for control of the malaria-carrying mosquito in an area where malaria is endemic. There will likely be some chemicals associated with domestic wastewater that will be allowed to adversely affect some part of the environment since the costs associated with their removal to the no-effect level may be sufficiently great to justify a small amount of environmental degradation. There will also be situations, especially in the early tiers of testing, in which the envelopes of estimated reliability for the no-effect concentration and the estimated environmental concentration overlap. Subsequent testing improves the precision of the estimates.

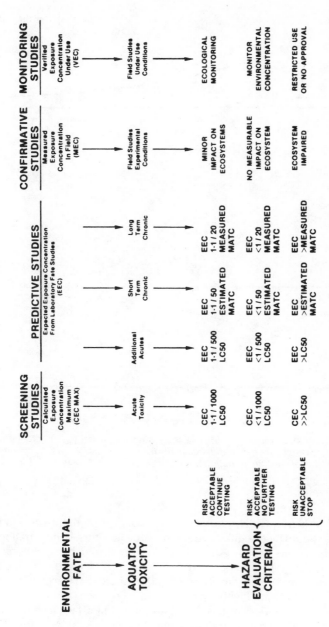

FIG. 1. Example of the tiered hazard assessment approach (from Kimerle et al. [5]). MATC, Maximum allowable toxic concentration.

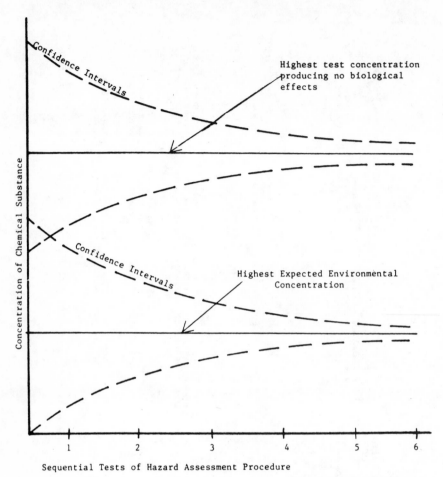

FIG. 2. *Diagrammatic representation of a sequential hazard assessment procedure demonstrating increasingly narrow confidence limits for estimates of no-biological-effect concentration and expected environmental concentration (from Cairns et al. [3]).*

Although it is frequently said that the toxicological component of a hazard assessment should be designed to determine the no-effect concentration, what is actually needed is a load-response relationship in which the environmental concentration-degree of impairment relationship is defined. The hazard assessment process then becomes one of finding the socially acceptable degree of impairment associated with a particular situation.

From the point of view of this discussion, it is important to note that estimated environmental concentration is one of the cornerstones of the decision-making process, the other being toxicity information. In the past, primary emphasis has been placed on developing aquatic toxicity data with little or no emphasis given to the development of comparable aquatic chemistry information. They both must be developed in tandem if a meaningful hazard assessment is to be made.

Industry should consider two other factors in this hazard assessment scheme as part of the decision-making process: the efficacy of the chemical for its intended use and its cost-profit margin. It is important that the cost of the hazard evaluation not be sufficiently great to cause industry to decide not to proceed toward the development of potentially useful chemicals. As discussed by Lee et al. (6), it must be recognized that of the tens of thousands of chemicals that enter the environment daily through the activity of humans, only a small number have caused chemical crises. Even for those crisis chemicals, there are significant questions about the actual environmental hazard that has been caused by them. It is extremely important that those responsible for development and use of hazard assessment schemes within both industry and governmental agencies not be so enamored of the scheme and the development of highly sophisticated test procedures as to forget the basic purpose of the scheme, namely, to screen chemicals for potential environmental hazard. It is important to note that the probability of any of the 500 or more new chemicals developed each year becoming the next PCB or Kepone is quite small. This situation has important implications for the development of appropriate methods for assessing biotransformation.

There are several key aspects of the various tiered hazard assessment schemes such as that presented in Fig. 1 which are pertinent to the design of procedures for assessing the biotransformability of a chemical. Depending on the design of the hazard assessment scheme, the first tier of testing can be as simple as examination of the structure of the chemical in order to estimate the tendency for biotransformation on the basis of known behavior of similar chemicals or functional groups within the chemical molecule. For example, certain types of functional groups are known to be readily attacked by aquatic microorganisms. This type of information can be used to make a preliminary assessment of the potential of microorganisms for causing an alteration in compound structure.

The next tier usually involves one or more laboratory assessments of biotransformation. For materials whose use would result in their being present in domestic wastewaters, laboratory activated sludge, anaerobic digestion, and river die-away tests would normally be conducted. These tests might be conducted to screen for overall biotransformation potential without trying to define the kinetics (rate) of reaction. Detailed kinetic data (the form of the rate expression and the rate constant) are needed only for biotransformations that take place at what would be considered moderate rates. Highly labile and highly resistant compounds can be handled in a different way. There is little or no need for accumulating detailed kinetic data for compounds with first-order biotransformation half-lives of months to years. If the testing for other transformations shows the compound to be highly persistent, then from an environmental hazard assessment point of view, these compounds can be considered essentially conservative and their estimated environmental concentration can be based on dilution-dispersion models.

For highly labile compounds which are biotransformed within a few minutes to a few hours, there is little need for kinetic data on the parent compound, since it would not be expected to be present for a sufficient amount of time to have an adverse effect on aquatic organisms, except possibly under spill conditions, or for contaminants associated with wastewater discharges or other highly concentrated inputs. Work needs to be done to provide guidance to those conducting biotransformation studies as part of a hazard assessment, for evaluating the rates of reaction above and below which there is need for detailed kinetic data for biotransformation reactions. These ranges will be dependent on the use of the chemical and, therefore, its rate and mode of entry into the environment and the overall level of biological activity that occurs in the aquatic environment. Consideration must be given to what are considered oligotrophic (nutrient poor) and eutrophic (nutrient rich) waters. This point will be discussed further below.

Any meaningful hazard assessment program must consider not only the parent compound, but also the potential environmental significance of the transformation products. There is a strong tendency to suggest that a "complete" analytical characterization of the parent compound(s) and transformation products be conducted as part of the early tiers of testing. This approach could easily result in increasing the cost of tests in early tiers to the point of causing some potential manufacturers to stop testing and discard the chemical. It is important to reemphasize that based on past experience, the likelihood of a transformation product causing significant environmental degradation is small. Therefore, procedures should be used which can estimate the potential for biotransformation but do not require complete analytical characterization of the compound and its transformation products. Of particular importance in this regard is the use of radiolabeled compounds. Often, a much less expensive evaluation of the biotransformation potential of a compound can be made by following the behavior of the radiolabel rather than developing specific analytical procedures for determining the concentration of the chemical and its transformation products in complex matrices that typically exist in aquatic systems. It is important to note, however, that in the higher levels of tiered testing, an evaluation should be made of the compound or compounds that contain the label after the compound has been exposed to conditions which would promote biotransformations. There have been instances where biotransformations have occurred which were not detected by radiolabel techniques.

Sometimes a bioassay designed to measure acute toxicity can be a useful substitute for an analytical characterization of transformation products in a hazard evaluation scheme. Since the primary concern in such a scheme is often toxicity to aquatic organisms, bioassays conducted on an "environmentally aged" sample can provide valuable information on the potential hazard of transformation products of the chemical.

Consideration should be given in the higher tiers to assessing the potential for biotransformation of chemical transformation products. It is highly conceivable that hydrolysis or photolysis of a biologically persistent chemical might convert it to a form that could be readily attacked by microorganisms. All biotransformation studies of persistent chemicals should be conducted in a way that provides for organism acclimation and cometabolism. Further, for chemicals which are readily taken up by higher organisms, consideration should be given to the probability of biotransformation within the liver or some other organ that would not occur in the presence of aquatic microorganisms. Generally, emphasis should be given to evaluation of biotransformation under aerobic conditions, but anaerobic transformations should be considered for persistent compounds in higher-level tiers.

For certain types of compounds such as chlorinated hydrocarbon pesticides, exposure to anaerobic conditions results in a dehalogenization which permits microbial transformation of the compounds under aerobic conditions. The environmental significance of such a phenomenon is not understood. Although known to occur in laboratories, it does not appear that this phenomenon is of particular importance in the environment; otherwise, there would not have been the worldwide accumulation of chlorinated hydrocarbon pesticides and PCBs in sediments. Whereas these reactions have not eliminated the environmental contamination by these chemicals, they may have minimized their adverse impacts. It is important that any laboratory testing of chemicals properly simulate the environment(s) they will encounter based on manufacture and use patterns. From an overall point of view, those responsible for development of procedures to assess biotransformation should give consideration to the methodology and guidance on how it should be used in a tiered hazard assessment scheme.

ESTIMATING ENVIRONMENTAL CONCENTRATION FOR NEW CHEMICALS

Since environmental fate and concentrations of a new chemical cannot be directly measured prior to manufacture and use of the chemical, they must be estimated. Baughman and Lassiter (1) and Branson (2) have discussed the development of mathematical models which can describe the expected concentrations in various components of the environment based on properties of the chemical, modes of input to the environment, and reactions that can occur in the environments with which it comes in contact. A schematic representation of this modeling approach is shown in Fig. 3.

Branson (2), in developing Fig. 3, did not include a number of chemical reactions such as biotransformations, oxidation-reduction, precipitation, complexation, etc., that may occur within aqueous systems and that would be important in influencing estimated environmental concentrations for a hazard assessment. The differential equation presented in Fig. 3 should be expanded to include terms for each of these reactions. In formulating the rate expression for biotransformation reactions, the typical approach used for biochemical oxygen demand (BOD) formulation, involving a first-order rate expression, should not be used. The rate of biotransformation of a compound is dependent on the enzyme activity within the system. Therefore, some measure of enzyme activity such as the concentration or organisms present that can bring about a certain type of transformation should be part of the rate expression. This means that the typical biotransformation rate expression should involve a second-order formulation in which the rate of reaction is dependent not only on the substrate concentration, but also on the enzyme activity within the sample.

The potential significance of the variety of reactions into which a chemical could enter in the environment, such as hydrolysis, photolysis, sorption, uptake by organisms, precipitation, and biotransformation, must be assessed either from knowledge of fundamental properties of the chemical or through laboratory experiment. The rates of those reactions determined to be of potential importance in defining the concentrations of the chemical in various

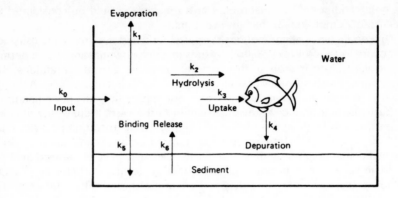

MATERIAL BALANCE EQUATION

$$V\frac{dCw}{dt} = k_0 - k_1ACw - k_2VCw - k_3FCw + k_4FC_f - k_5SCw + k_6SC_s$$

Input | Evaporation | Degradation (Hydrolysis) | Fish Uptake | Fish Depuration | Sediment Binding | Sediment Release

Where V = Volume of water, ml

A = Surface area, cm²

F = Fish mass, gm

S = Sediment mass, gm

Cw = Concentration of chemical in water

k = Rate constant

C_f = Concentrations of chemical in fish

C_s = Concentration of chemical in sediment

FIG. 3. *Example of environmental chemistry-fate model for environmental hazard assessment (from Branson [2]).*

parts of the environment (fate) must be experimentally estimated. Based on a mass balance approach such as that depicted in Fig. 3, a series of differential equations is then formulated describing the rates of the pertinent reactions. It is important to note that since this type of model is based on the characteristics of the chemical rather than on a particular system, it has a potentially wider range of applicability than "dynamic" models tuned to the behavior of a chemical in a particular system. Committee E-35 of the American Society for Testing and Materials (section 21.02) is actively involved in the development of mathematical environmental chemistry-fate models for use with new and expanded-use chemicals as part of an environmental hazard assessment.

Without environmental chemistry-fate modeling, environmental hazard assessment would have to be based on a worst-case environmental concentration, in which only dilution is considered. This approach would be highly restrictive for development of new chemicals since it would not make allowances for the wide variety of transformations that occur in environmental systems. It is therefore important in an environmental hazard assessment program to be able to reliably estimate the rate and extent of chemical-biochemical transformations that may occur in environmental systems, since this will allow a much greater use of a chemical, i.e., greater environmental load, than if the only mechanism for concentration reduction were the physical processes of mixing-dilution.

The other modeling approach that has been used to estimate environmental fate for new chemicals is microcosms in which an attempt is made to simulate an environmental system in the laboratory. It is generally agreed by those knowledgeable in environmental chemistry-fate modeling that the microcosm approach, using microcosms of the type that have been used for pesticide screening, has little or no predictive capability. The results of a microcosm experiment are dependent upon the types of components and their relative amounts (i.e., the numbers of organisms, amount of substrate, etc.). Unless great care is used to properly simulate the environment, the microcosm approach can lead to erroneous conclusions.

Microcosms do have a potentially important role in environmental hazard assessment for new chemicals. This role is in the verification of the mathematical model. It is believed that every environmental chemistry-fate mathematical model should be evaluated with respect to its ability to predict the concentrations of the chemical of concern in the various components of a laboratory microcosm. If the mathematical model does not reliably predict contaminant behavior in a simple laboratory microcosm, then its potential reliability as a tool to estimate environmental concentration in the natural environment will likely be extremely limited. One of the greatest needs that exists today in developing estimates of environmental concentration as part of a hazard assessment is verification of their predictive capability for existing chemicals already in commercial use. Studies of this type will be extremely important in providing guidance on further model development. Models of the type developed by Baughman and Lassiter (1), Branson (2), and the American Society for Testing and Materials are oversimplifications of the actual environmental behavior of chemicals. Although at this time it is not known how well these models predict environmental behavior for a variety of chemical types in different environments, verification studies will provide the kinds of information needed to determine whether future model development will require greater emphasis on expanding the models to include greater numbers of types of reactions or refinement of the test procedures, such as for biotransformation, in order to provide more reliable data input to the model.

With the manufacture and initial sale of a new chemical, studies should be conducted to verify the environmental chemistry-fate models used in evaluating the environmental hazard of the compound. Many industries use a test marketing approach in which sale will be restricted to a limited area. Associated with any test marketing designed to evaluate consumer acceptance should be studies conducted to verify the environmental chemistry-fate model developed for the chemical. If major discrepancies between predicted and actual concentrations are found, then modifications of the model should be made to improve its predictability.

In addition to the need for environmental concentration data for new chemicals, there is also a need for data of this type for existing chemicals which have a potential for having expanded or new uses and in connection with the development of meaningful discharge limitations. With respect to the latter area, increased attention is being given to establishing limitations for contaminant input based on actual environmental impact. This approach will require the use of environmental chemistry-fate models of the same type being developed for new chemicals in order to predict the concentrations and forms of the contaminant of concern in various parts of the receiving waters under altered load conditions.

Historically, the approach of the United States and other countries to the control of chemicals in the aquatic environment has been to allow the manufacture and use of a chemical until a water quality or other environmental quality problem is found or implicated to be associated with it. Beginning in the 1960s, the control of chemical contaminants from a point source has been based on the degree of contaminant removal that can be readily achieved by existing technology. U.S. Public Law 92-500 mandates this approach through the "best practicable" and "best available" treatment requirements for removing chemicals from point sources. Although contaminant control programs are now usually based on a fixed percentage removal of materials such as suspended solids and BOD, the emphasis in the 1980s will shift to a hazard assessment approach. The degree of removal required for "hazardous chemicals" from a source will be determined by a stream classification system in which limits based on the desired water quality in the receiving waters will be set.

Public Law 92-500 also mandates that the U.S. Environmental Protection Agency (EPA) develop water quality criteria that will serve as a basis for development of state water quality standards. It has become clear that without qualifications, the EPA "Red Book" criteria of July 1976 are not suitable for adoption as numeric state water quality standards. These criteria were developed on the basis of a worst-case situation; they were generally developed by exposure of the test organisms to 100% available forms of the chemicals for "chronic" durations (or were extrapolated to chronic exposure). Chemicals exist in aquatic systems in a variety of chemical forms, some of which are not available at a sufficient rate and extent to be adverse to the organisms or higher-trophic-level organisms. In many systems, the concentrations of the various chemical forms are highly variable. Water quality standards must therefore take into account the concentration of available forms and duration of exposure relationships that exist in aquatic systems.

Recently, Lee and Jones (8) completed a comprehensive review of sets of International Field Year for the Great Lakes (IFYGL) concentration data for selected contaminants in the nearshore waters of Lake Ontario. Using this IFYGL data base, they found that concentrations for each of the contaminants investigated were highly variable over time, with excursions considerably above the EPA water quality criteria and Water Quality Agreement Objectives selected by the United States and Canada for the Great Lakes (4).

To assess the significance of the "excessive" (as judged by the "Red Book" criteria) concentrations of a chemical with regard to impairment of beneficial uses of water, including aquatic life, an environmental hazard assessment of the type that is being developed under the requirements of the EPA TSCA must be made. Toxicological information on the concentration of available forms of the contaminant and on the duration of exposure of organisms to the available forms necessary to produce an adverse effect must be used with an environmental chemistry-fate modeling approach to predict environmental impairment that will result from the input of a particular chemical into a specific body of water. This type of information needs to be examined in light of engineering cost data for control of the contaminant to various levels in order to establish the socially desired level of water quality—acceptable water quality impairment for a particular body of water. This approach is being developed within the EPA for examination of the environmental impact of complex mixtures of chemical wastes. The adoption of this approach throughout the water quality control field will lead to much more cost-effective, technologically valid, and ecologically sound water quality management than is being practiced today.

ADDITIONAL CONSIDERATIONS

There are several aspects of the development of biotransformation data that need particular attention. One of the most significant problems facing those concerned with the development of biotransformation data for an environmental hazard assessment is relating organism activity levels that occur in the laboratory to levels encountered in the aquatic environment. As discussed previously, one of the key components of an environmental hazard assessment is an estimate of the environmental concentration of the chemical. This will require, for compounds that have "moderate" rates of transformation, detailed kinetic data which will provide an estimate of how rapidly the biotransformation will occur in various parts of the aquatic environment. This rate will be dependent on the numbers, types, and, most importantly, the activity of certain types of aquatic organisms. At the present time, procedures for developing this type of information are generally lacking. It is relatively easy to conduct laboratory studies which will determine whether or not a compound will be biotransformed. It is difficult to conduct these studies so as to develop kinetic data applicable to estimating the rate of biotransformation in aquatic systems. Those responsible for developing biotransformation procedures must give particular attention to developing approaches which can be used to translate laboratory rate data into field rate data. As noted above, this will require a reliable estimate of enzyme activity.

There is immediate need for a substantial research program designed to provide information in the field of environmental microbiology. Studies should be conducted to determine the rates of biotransformation in aquatic systems and especially the factors controlling these rates. Once the primary factors are understood, then emphasis should be placed on developing the types of information necessary to translate environmental biotransformation data from the field to laboratory situations and vice versa.

Two of the parameters that should be considered as measures of environmental enzyme activity of aquatic microorganisms are the numbers and types of microorganisms in various aquatic environments determined by plating techniques and biochemical activity determined by enzymatic tests. It is important to note that it is not the concentration of microorganisms that is of major importance, but rather their activity. It may, however, be possible to empirically relate organism concentration data to activity. It also may be of value to have an estimate of overall aquatic microorganism activity levels as measured by diel changes in dissolved oxygen, adenosine triphosphate, and ribonucleic acid. One or more of these parameters might be empirically correlated to the overall rate at which certain types of compounds may be biotransformed in aquatic systems. Further, there may be a general relationship between the classical oligotrophic-eutrophic classification of water bodies and the biotransformation potential of aquatic systems for certain types of compounds. As discussed by Lee and Hoadley (7), it appears that the bacterial populations of aquatic systems are closely correlated with phytoplankton production. Vollenweider (11), Rast and Lee (10), and Lee et al. (9) have shown that for many water bodies, there is a remarkably good correlation between the phosphorus load to the water body (as normalized by the water body's mean depth and hydraulic residence time) and the algal biomass produced within the water body. It is conceivable that a relatively simple parameter such as planktonic algal chlorophyll or, for phosphorus-limited water bodies, the normalized phosphorus load could be a sufficiently reliable indicator of overall organism activity levels to be useful in characterizing aquatic environment biotransformation potential as part of an environmental hazard assessment for new or expanded-use chemicals.

Another area that needs particular attention is the role of natural water particulate matter in influencing the rates of biotransformation in aquatic systems. Many of the chemicals of greatest interest as potential environmental hazards tend to be strongly sorbed by natural water particulate matter. This sorption could greatly influence the rate at which aquatic microorganisms transform the chemical. The development of biotransformation test protocols should include an evaluation of the potential significance of sorption reactions in influencing the rates of reactions in aquatic systems. This work should include the influence of not only large (>1-μm diameter) particles, but also of colloidal particles.

In the mid-1960s, the persistence of alkyl benzene sulfonate, a detergent surfactant, in the aquatic environment triggered a substantial effort toward characterization and understanding of biotransformation in aquatic systems. This ultimately resulted in the development of a biodegradable surfactant for use in detergent formulations. We need now to initiate a level of aquatic biotransformation research which would be considerably greater in funding than that available during the peak of the alkyl benzene sulfonate activity. Without this level of funding, one of the basic foundations (biotransformation information) of environmental hazard assessment approaches being used by industry and government will not be developed to its full potential. It is imperative that this area receive much greater attention than it has in the past if meaningful hazard assessments are to be made for new or expanded-use chemicals.

Guidance for this work can be obtained from the many years of study associated with the impact of domestic and industrial wastewater on the oxygen resources of aquatic systems. The BOD test is a biotransformation test designed to estimate, based on laboratory conditions, the rate and extent of biotransformation (i.e., oxygen utilization) that will occur in aquatic systems. This test has been perfected to such a degree that it is a reliable predictor of aquatic biotransformation for oxygen-demanding wastewater components. As noted above, however, the major deficiency of this test for use in hazard assessment programs is that it does not include an estimate of the dependence of test results on enzyme activity levels. Studies should be done to assess bacterial population activity within the test system and in aquatic systems to properly formulate the relationship between laboratory-derived BOD data and biotransformation rate in a variety of aquatic systems.

LITERATURE CITED

1. **Baughman, G. L., and R. R. Lassiter.** 1978. Prediction of environmental pollutant concentration, p. 35–54. *In* J. Cairns, K. L. Dickson, and A. W. Maki (ed.), Estimating the hazard of chemical substances to aquatic life. ASTM Special Technical Publ. 657. American Society for Testing and Materials, Philadelphia.
2. **Branson, D. R.** 1978. Predicting the fate of chemicals in the aquatic environment from laboratory data. *In* J. Cairns, K. L. Dickson, and A. W. Maki (ed.), Estimating the hazard of chemical substances to aquatic life. ASTM Special Technical Publ. 657. American Society for Testing and Materials, Philadelphia.
3. **Cairns, J., K. L. Dickson, and A. W. Maki (ed.).** 1978. Estimating the hazard of chemical substances to aquatic life. ASTM Special Technical Publ. 657. American Society for Testing and Materials, Philadelphia.
4. **International Joint Commission.** 1978. Great Lakes Water Quality Agreement of 1978. Agreement with annexes and terms of reference between the U.S.A. and Canada. International Joint Commission, Windsor, Ontario.
5. **Kimerle, R. A., W. E. Gledhill, and G. I. Levinskas.** 1978. Environmental safety assessment of new materials. *In* J. Cairns, K. L. Dickson, and A. W. Maki (ed.), Estimating the hazard of chemical substances to aquatic life. ASTM Special Technical Publ. 657. American Society for Testing and Materials, Philadelphia.
6. **Lee, G. F., T. Heinemann, and R. A. Jones.** 1979. Comments on March 16, 1979, *Federal Register.* Part IV. Toxic substances control—discussion of premanufacture testing policy and technical issues. Request for comments. Occasional Paper no. 40. Environmental Engineering, Colorado State University, Fort Collins.
7. **Lee, G. F., and A. F. Hoadley.** 1967. Biological activity in relation to the chemical equilibrium composition of natural water. Adv. Chem. Ser. **67:**319–338.
8. **Lee, G. F., and R. A. Jones.** 1978. Water quality characteristics of the U.S. waters of Lake Ontario during the IFYGL and modeling contaminant load–water quality response relationships in the nearshore waters of the Great Lakes. National Oceanographic and Atmospheric Administration Great Lakes Research Laboratory, Ann Arbor, Mich.
9. **Lee, G. F., W. Rast, and R. A. Jones.** 1978. Eutrophication of water bodies: insights for an age-old problem. Environ. Sci. Technol. **12:**900–908.
10. **Rast, W., and G. F. Lee.** 1978. Summary analysis of the North American (US portion) OECD eutrophication project: nutrient loading–lake response relationships and trophic state indices. EPA 600/3-78-008. U.S. Environmental Protection Agency, Corvallis, Ore.
11. **Vollenweider, R. A.** 1976. Advances in defining critical loading levels for phosphorus in lake eutrophication. Mem. Ist. Ital. Idrobiol. **33:**53–83.

Synopsis of Discussion Session: Environmental Concentration—Background

G. LOEWENGART AND G. F. LEE

Estimating the environmental concentration of a material requires a reliable estimate of the amount of material released and the rate of input to the environmental compartments under consideration and an understanding of the dissipative and assimilative processes that will modify this concentration. In general, the modifying and mitigating processes are categorized as chemical, physical, and biological. Although all of these processes are important, the focus of this workshop was on biotransformation of chemicals in the aquatic environment. The purpose of this workshop was not to advance the state of the art of the science of biotransformation, but to critically assess it, determine how it affects the environmental concentration, and apply that information to the generalized hazard assessment process.

The basic principle developed by the two previous workshops dealing with aquatic hazard assessment (1, 2) was that in performing a hazard evaluation for a chemical in the aquatic environment it is necessary to quantify the risk associated with exposure to a chemical, and this leads to the concept of risk as the inverse of the margin of safety. Determination of the margin of safety requires an estimation of the no-observed-effect concentration (NOEC) and the estimated environmental concentration (EEC).

To a great extent, emphasis to date has been on obtaining an increasingly accurate estimate of the NOEC. If reliable hazard assessments are to be made, it is essential that the EEC also be accurately developed to complement the NOEC. Expressed in another way, in the process of evaluating the environmental safety of a chemical, we need to be concerned with both the effects of the chemical on the environment (NOEC) and the effect of the environment on the chemical (EEC).

APPLICATION OF EEC TO ENVIRONMENTAL HAZARD ASSESSMENT

Several points are relevant to the application of the data used to obtain the EEC in an environmental hazard assessment:

1. An estimate of the environmental concentration is necessary in virtually all environmental hazard assessment procedures, whether or not these are sequential, tiered assessment programs.
2. The concept applies not only to new chemicals, to be considered under the Toxic Substances Control Act (TSCA), but also to any chemical, mixture, effluent, and even solid wastes which can reach the environmental compartment under consideration.
3. The laboratory data developed and estimates derived therefrom should have application to the "real world" environment.

4. It is essential to clearly define the objective and develop an approach for estimation of environmental concentration to satisfy the objective with the necessary degree of precision and accuracy.

The nature of the chemical, the production volume, and the release rate pattern of manufacture and use will all affect the approach used in estimating the environmental concentration. Obviously, varying degrees of detail and precision of the data developed to generate the EEC will be required to tailor the end product to the objective. For example, in some cases a complex, lengthy research approach will be appropriate, whereas in others a simple management decision tool is required to screen a large number of chemicals.

GENERAL COMMENTS

The estimation of environmental concentration can be made only if information is developed about the rates and extent of the biological transformation. It is important to note here that single-point biochemical oxygen demand values may be considered as crude rate measures of microbial transformation in aquatic systems.

It is also important to avoid the simplistic impulse to propose in-depth study of every chemical. That approach is not practical or logical with limited scientific and financial resources. A mechanism must be developed whereby highly degradable and highly resistant chemicals are identified early and clearly. Little additional work will be necessary for most chemicals at these extremes. Decisions can then be made for those intermediate-range chemicals which may need more extensive studies to estimate their environmental concentrations accurately.

A word of caution—single tests, highly specific systems, and complex all-inclusive ecosystem models, although useful in selected circumstances, have limited applicability to estimates of expected environmental concentrations as related to the environmental hazard assessment process. Some workers in the field have methods and approaches in which, by reason of constant use or familiarity, they have developed vested interests. These methods are then rightly pushed to their limits, but, unfortunately, sometimes beyond their appropriate application.

FUTURE PERSPECTIVES

Experiences of the recent past suggest that it is likely that fewer materials which are both highly persistent and bioaccumulative will be introduced into the environment. In addition, with the introduction of widespread application of hazard assessments to new chemicals, materials that are highly toxic and those of a very low order of toxicity relative to their expected environmental concentrations will be identified and evaluated early and relatively easily. If this proves to be the case, the focal point in the future will be the environmental effects of materials of intermediate toxicity, persistence, and bioaccumulation in addition to concern for widespread contamination of the environment by persistent chemicals. Thus, the area at the boundary of and just past the mixing zone associated with the point of entry of a chemical into a receiving water may become the environmental area at risk. In that area, if effects occur, they will be manifested a few hours to several days after contaminant release into the environment. The data required to evaluate the significance of the chemical in this area of highest environmental concentration will be the relationships between concentration-duration of exposure and toxic effect-concentration.

LITERATURE CITED

1. **Cairns, J., K. L. Dickson, and A. W. Maki (ed.).** 1978. Estimating the hazard of chemical substances to aquatic life. ASTM Special Technical Publ. 657. American Society for Testing and Materials, Philadelphia.
2. **Dickson, K. L., A. W. Maki, and J. Cairns (ed.).** 1979. Analyzing the hazard evaluation process. American Fisheries Society, Bethesda, Md.

Chapter 3. EVALUATION OF METHODS

Approaches to Estimating Microbial Degradation of Chemical Contaminants in Freshwater Ecosystems

B. THOMAS JOHNSON

Columbia National Fisheries Research Laboratory, U.S. Department of the Interior, Fish and Wildlife Service, Columbia, Missouri 65201

A review of microbial methods for investigating the biodegradability of chemical contaminants revealed omissions in experimental design that tended to negate the usefulness of the methods for estimating the persistency of chemicals in natural environments. Emphasis was on chemical analysis and identification of degradation products and pathways; little consideration was given to the biological part of the degradation test. An integrated approach to the investigation of microbial degradation in aquatic ecosystems is proposed in the hydrosoil biodegradation test. Estimation of the persistency of a xenobiotic in freshwater ecosystems requires consideration of contaminant sorption and concentration, hydrosoil minerals, biogenic and xenobiotic organic content, and microbial activity.

The persistency of a chemical contaminant in the environment is a key factor in determining the magnitude of the stress of the contaminant on the indigenous flora and fauna of the aquatic community. Persistence is herein operationally defined as a chemical's resistance to biotic, abiotic, or photochemical degradation. Acute and chronic toxicity to fish and invertebrates, perturbation of biogeochemical cycles, and food chain bioaccumulation are directly influenced by the environmental persistence of the contaminant. Assessment of a chemical's biodegradability, before its commercial development and release into the environment, is an integral part of the Toxic Substances Control Act of 1976. An assessment of the persistency of a chemical contaminant is without question an important factor in estimating the potential environmental hazard of that chemical.

There are many varied ways to investigate microbial degradation in soil and water (2, 7). The most frequently used tests (2, 7) involve river die-away, biochemical demand, shake culture, model ecosystem, activated sludge, anaerobic systems, soil, and pure cultures. Although the literature abounds with methods of investigating microbial degradation of chemical contaminants, most are not suitable for environmental assessment.

It is difficult if not impossible to estimate, much less compare, the environmental fate of most synthetic organic contaminants. I would like to address this question: Why are these protocols inadequate? Because of the scope of the topic and my interest, I will restrict my comments to the freshwater environment. I think we need first to consider the protocol. What is a microbial chemical contaminant test? What is its design? What questions are asked of the test?

A review of the literature (2, 7) revealed much about the experimental design of most microbial studies concerned with synthetic organic chemical biodegradation; it also revealed the pitfalls that invalidate the design as a useful estimate of environmental persistency.

The basic experimental design of the degradation test consists of three parts: (i) the chemical (contaminant), (ii) the biological (the microbial decomposer), and (iii) the analytical (the test). Two questions are generally asked: (i) Do microorganisms (usually bacteria) degrade (metabolize or cometabolize) the chemical contaminant? (ii) What are the products (or intermediates) of degradation—and the probable pathway? Tests include a wide range of approaches, but those in which the ^{14}C-labeled contaminant is used are the most reliable. Evidence of degradation is monitored by $^{14}CO_2$ radiorespirometry, thin-layer chromatography, and autoradiography (gas chromatography, radioactivity monitoring, and high-pressure liquid chromatography are available and frequently used). The major emphasis in most protocols is on the "analytical" segment of the experimental design; only token consideration is given to the "chemical" or "biological" components.

A review of degradation protocols (2, 7) designed for laboratory use revealed a pattern of omissions that tended to negate the usefulness of the test as a predictive tool in estimating environmental persistency of a chemical contaminant. I have listed a number of questions divided into three groups paralleling the experimental design of a degradation test: chemical, biological, and analytical. They are not listed in any particular priority.

The chemical questions are as follows: What is the rationale for the concentration of the contaminant in the test? Is it based on use pattern or geographical distribution? What is the water solubility of the contaminant? Is the contaminant biologically available to the microbial decomposer? What is known about its sorptive capabilities? To what general chemical group does the contaminant belong? Can predictions be made concerning possible degradation products or pathways?

These are frequently asked biological questions: What is the microbial biomass in the test? How is it measured? How "typical" is the microbial sample? Does biomass type (sample source) influence degradation? What is the physiological profile of the microbes present? What is the limnological-geological profile of the sample (hydrosoil)? Do biogenic and xenobiotic compounds (and concentrations) influence degradation?

A number of analytical questions are important: What is the rationale for specific physical properties (pH, Eh, temperature, dissolved oxygen) used? How relevant are they to environmental conditions where the contaminant might enter? What are the "routine" physical variables used and their rationale? Is an accurate ^{14}C-labeled contaminant balance (input-output) measured? What is the accountability (^{14}C) of the test? What statistical treatment is applied to the sample? Is the experiment repeated and, if so, how many times? How many replicates are made and what determines their number? What is the variance of the sample? What variance is acceptable? Is the test a part of an integrated interdisciplinary study of a chemical contaminant?

These questions should be considered as I design a method for estimating the biodegradability of a synthetic organic compound in a freshwater ecosystem. For discussion, I have selected a hydrosoil biodegradation test (HBT) as a vehicle for examining a biodegradation method. This presentation should not be considered an exhaustive survey of the literature; however, I have noted

the weaknesses in many other methods and have developed appropriate measures to correct them.

Parenthetically, I believe a few comments are in order. Certainly, a hydrosoil
degradation study can be done with a grab sample from a pond or lake and
a ^{14}C-labeled (or even nonlabeled) xenobiotic by testing a few variables such
as temperature, dissolved oxygen, and chemical concentration. But one may
question just how far these results can be extrapolated to the "real world."
How representative is the "sample"? The use of pure culture or enriched cultures may indicate the probable microbial degradability of the test chemical,
but how does microbial biomass (or physiological type) or physical sorption
(or other environmental factors known or unknown) influence the degradative
process? Recent pure culture studies with polychlorinated biphenyls (PCBs)
(3) and activated-sludge investigations with phthalates (5, 17) suggest a number
of serious pitfalls associated with one-dimensional testing. One may accept
certain tenets of the concept of universality of microbial degradation in the
laboratory that are inherent in the pure culture approach, but do we really
know enough about microbial ecology to make the sweeping statement that
"microbial degradability" connotes—particularly where the conclusion will
have a far-reaching impact on water quality and fishery resources?

I have considered a number of factors that may be important in the environmental degradation of xenobiotics in freshwater. I have also considered a
number of investigative approaches to test the effect of these factors in the
laboratory (although some may view this exercise as a scientific "fishing ex-
pedition," I believe the rationale will soon become evident). These investigative approaches may or may not survive the economic and scientific crucible
of critical scientific investigation. I hope, however, that this paper will stimulate
interest in the total microbial biodegradation test—chemical, biological, and
analytical.

Briefly, a word about HBT. I developed the HBT at the Columbia National
Fisheries Research Laboratory to investigate the degradation of xenobiotics in
natural freshwater ecosystems. In a water-sediment (hydrosoil) microcosm-
type system, ^{14}C-labeled contaminants and the natural microbiota of hydrosoil
are used. Hydrosoil is designated as the zone rich in microbes, particularly
bacterial heterotrophs, in the sediment at the water interface. The test is intended to be both a qualitative and a quantitative measurement of microbial
degradation of aquatic chemical contaminants; these data are used in a predictive model for assessing the impact of xenobiotic stress on a freshwater
ecosystem. The formal objectives of HBT are four: (i) to determine the biodegradability of xenobiotics by the natural microbiota in freshwater hydrosoil,
(ii) to determine the products and possible pathways of the microbial degradation, (iii) to determine the chemical, physical, and biological factors that
influence the microbial degradation process in freshwater hydrosoil, and (iv)
to develop relevant and valid data that can be extrapolated to natural communities (the real world) for estimating the potential environmental stress of
chemical contaminants. The protocol of HBT is divided into three parts: chemical, biological, and analytical. I will discuss each section, its rationale, the
expected results, and some specific procedures used. I have selected high-
lights that I believe are important to a relevant and valid experimental design.

Basic information concerning the chemical structure and physical properties of the xenobiotic is an essential and integral part of a relevant test protocol. The presence of a chemical contaminant in a freshwater community does not in itself imply its availability to the microbiota for degradation or mineralization. The contaminant must first be biologically available (15). The chemical can be bound to clay particles or organic fragments and relatively inaccessible to microbial enzymes. To estimate the biological availability of a xenobiotic in the hydrosoil, one must have basic information about the chemical: (i) its solubility in water, (ii) its sorption to hydrosoil, (iii) its volatility, and (iv) its solubility in organic solvents.

I developed the hydrosoil sorption test, modified after Lee and Mariani (12), to determine the potential biological availability of the xenobiotic in hydrosoil. The test procedure follows. Fill each of six 125-ml Erlenmeyer flasks with 10 g of hydrosoil and 90 ml of lake water and ^{14}C-labeled contaminant. Place half of the samples on a magnetic stirrer for 30 min to thoroughly mix the soil, water, and ^{14}C-labeled contaminant; gently stir only the lake water in the remaining samples for a few minutes with a glass rod. Immediately place the flasks in an incubator and maintain them at 22°C for 24 h. Carefully siphon off the supernatant lake water and analyze it by a liqud scintillation method (9). Add an equal volume of water to replace the lake water. Minimize the disturbance of the hydrosoil while removing and replacing water in the samples. Sample both the water and the soil at 7 days. Extract the samples with a mixture of methanol-toluene, methanol-ethyl acetate, or methanol-dichloromethane; choose an appropriate solvent. Adjust the pH to 1 to 2 and 11 to 12 during extraction. Mix 1 part hydrosoil with 4 parts anhydrous sodium sulfate (weight/weight) and extract in a glass column. After two extractions, burn 1.0 g of hydrosoil in a bio-oxidizer oven (Harvey Instrument Co., Hillsdale, N.Y.) at 700°C and trap the ^{14}CO$_2$ with an organic amine (Harvey Instrument Co.). Quantify the radioactivity of both the soil and the ^{14}CO$_2$ in a liquid scintillation counter. Account for all the ^{14}C-labeled contaminant. Prepare an index based on a comparison of the hydrosoil sorption of the test contaminant from water with that of p,p'-DDT (dichlorodiphenyltrichloroethane) from water under similar conditions. Give DDT an arbitrary value of 100 in the hydrosoil sorption index.

The concentration of the xenobiotic under investigation should be relevant to potential environmental exposure. Proposed usage patterns are a good point of reference. Concentrations between 1 μg/kg and 10 mg/kg are routinely used in the hydrosoil test. Consideration of either the high or the low influence of the chemical concentration on the degradation process was also made (1).

Preincubation with the test chemical could be considered, particularly when the use pattern suggests that the hydrosoil will be chronically exposed to the contaminant. Length of exposure is best considered after preliminary data concerning the degradability of the xenobiotic in the hydrosoil have been gathered. Preincubation with a readily degradable compound is for about 7 days (at the similar concentration of treatment), whereas a "recalcitrant" compound, such as Kepone, is preincubated for at least 30 days.

The type of hydrosoil samples taken are essentially tailored to the specific chemical contaminant, whether the soil is from a lentic or a lotic source, littoral or profundal, and whether water is eutrophic or oligotrophic, epilimnitic or hypolimnitic. Use patterns and geographical distribution of the chemical provide a good basis for this decision.

Geological, limnological, and biological characterizations are essential measurements and a necessary prerequisite for valid biodegradation testing and evaluations.

All soils were analyzed for total organic carbon, total organic nitrogen, clay fraction, Ca, Mg, K, Na, Cl, HCO_3, SO_4, and total dissolved solids. The total concentration of dissolved substances or minerals in natural hydrosoil is a useful property in "describing the chemical density as a fitness factor" (16) for biological activity in the aquatic community.

The history of possible xenobiotic exposure to the hydrosoil was also gathered. The soil samples were analyzed for residues of pesticides and industrial by-products, particularly organochlorine compounds.

Assessment of microbial activity in hydrosoil was studied by determination of microbial biomass, cell constituent, respirometry, and substrate accumulation. Most of these methods have been adequately described in the literature. Jannasch (8) reviewed various approaches for the determination of microbial activity in water.

The total number of heterotrophic bacteria in hydrosoil was determined by the most-probable-number (MPN) method (4). The MPN values were determined by growth on either casein-peptone-starch medium (10) or nutrient agar. Five 10-fold dilutions were made of hydrosoil; each dilution was represented by five replicates. After the plates were incubated at 22°C for 4 days, cell numbers were determined from an MPN table (4).

I chose the direct-count method of Hobbie et al. (6), using acridine orange stain and an epifluorescence microscope, to count hydrosoil bacteria.

Adenosine triphosphate content of the hydrosoil was measured by the methods of Karl and LaRock (11) and Paul and Johnson (14). The DuPont Biometer system (DuPont Instruments, Wilmington, Del.) was used to measure emission from the luciferin-luciferase reaction.

A Gilson differential respirometer (Gilson Medical Electronics, Inc., Middleton, Wis.) was used to measure oxygen consumption at 28°C. Each reaction vessel contained 1.5 g (wet weight) of hydrosoil overlaid with 4.5 ml of lake water. The center well contained 0.2 ml of 10% KOH; a small fluted sheet of filter paper was used to increase the CO_2 absorbance by the KOH. After temperature and pressure equilibration, 1.0 ml of 0.25% glucose was tipped from the side arm. Oxygen consumption of the hydrosoil was measured for 2 h. All experiments were performed with five replicates. Hydrosoil in the control flasks was autoclave killed and treated with 1% $HgCl_2$. Oxygen consumption was measured per dry weight of hydrosoil.

The dehydrogenase technique (13) for measuring the metabolic activity of hydrosoil under anaerobic conditions was modified to use a 6-h incubation period at 28°C with 1% peptone as the electron-donating substrate. The test is based on the reduction of TTC (2,3,5-triphenyltetrazolium chloride) to a red ethanol-soluble pigment, formazan. One gram of hydrosoil was dispensed into 50-ml glass centrifuge tubes and overlaid with 11 ml of lake water. The hydrosoil was preincubated in the tubes for 24 h at 22°C in a stainless-steel anaerobic container (Torsion Balance Co., Clifton, N.J.), evacuated under vacuum, and overlaid with nitrogen gas. After the anaerobic preincubation, 3.0 ml of 1% peptone, 2.0 ml of 1% TTC, and 30 ml of tris(hydroxymethyl)aminomethane buffer (pH 8.4) were added to the hydrosoil and mixed thoroughly on a Vortex mixer. The tubes were returned to the anaerobic jar, evacuated under vacuum, overlaid with nitrogen, and incubated for 6 h at 28°C. After incubation, the tubes were immediately extracted with 20 ml of 95% ethanol for 30 min with vigorous mixing on a reciprocal shaker. The tubes were centrifuged at about 5,000 × g for 10 min to separate the ethanol containing the formazan. Duplicate 1.0-ml samples of the red ethanolic supernatant were measured at 483 nm with a Spectronic 100 single-beam spectrophotometer (Bausch and Lomb) equipped with a micro flow-through cell. These values were compared with a formazan standard curve and reported as formazan per unit of dried hydrosoil. All experiments are replicates of five samples.

Wright (20) has proposed an interesting approach to the measurement of heterotrophic bacteria in natural waters. His concept of "specific activity" that attempts to give "quantitative expression to the relationship between heterotrophic activity and bacterial direct counts" (20) may be a very valuable tool in evaluation of hydrosoil degradation data. His index is essentially a ratio of a physiological measurement that varies directly with bacteria from a natural hydrosoil sample. Wright measures heterotrophic activity with ^{14}C-labeled substrate and directly counts bacteria with the acridine orange and epifluorescence microscope technique. We at the Columbia National Fisheries Research Laboratory have used his methods; in preliminary experiments with freshwater hydrosoils, the results seem promising. The specific activity concept may also prove to be valuable in evaluating xenobiotic degradation data obtained from different hydrosoil sources. It is worthy of serious consideration in any biodegradation protocol.

The hydrosoil inoculated with the ^{14}C-labeled contaminant is incubated under controlled laboratory conditions. Both qualitative and quantitative measurements of the hydrosoil and water are made during and after incubation. Possible variables for routine experiments are pH, temperature, dissolved oxygen (aerobic or anaerobic conditions), incubation times, and concentrations of the contaminant. A scheme for the cleanup, fractionation, analysis, and identification of the chemical contaminant is outlined in Fig. 2. This analytical flow sheet is based on a concept of complete accountability of the ^{14}C-labeled contaminant, a complete accounting of input, degradation, volatility, and sorption of the chemical during the biodegradation test. Extraction procedures and chromatographic methods used are tailored to the specific xenobiotic. Under anaerobic conditions strict ^{14}C-labeled compound accountability has not been satisfactorily worked out. Methanogenic bacteria indigenous to hydrosoil will use $^{14}CO_2$ (released during degradation of the ^{14}C-labeled test compound by heterotrophic bacteria) as a carbon source, reducing the amount of radiolabeled $^{14}CO_2$ trapped and consequently causing an underestimation of anaerobic degradation of the test compound. Recent work of Zehnder et al. (21) in which they measured radioactive methane with conventional liquid scintillation spectrometry shows promise as a possible answer to this problem.

Sample replication is determined by the dispersion of data from the sample mean. As a rule all experimental samples are replicated in triplicate; however, if the coefficient of variation (mean divided by standard deviation × 100) exceeds 20%, the sample size is increased to five. All data are expressed as mean ± standard deviation. All degradation rate studies are subjected to the regression analysis of Snedecor and Cochran (18).

An example of a hydrosoil test follows. The hydrosoil sample was obtained from a slightly eutrophic lake. The samples were incubated under aerobic and anaerobic conditions at 22°C for 28 days.

Ten-gram (wt/wt) portions of hydrosoil were placed in 125-ml Erlenmeyer flasks and covered with 90 ml of lake water. The lake water was collected with the hydrosoil and was the natural water column over the hydrosoil. The flask contents were then inoculated with the ¹⁴C-labeled contaminant, usually in an acetone or ethanol carrier. The carrier volume was limited to 0.1% or less of the total flask volume to avoid possible inhibition (toxic or competitive) of microbial populations. A ¹⁴C-labeled contaminant concentration of 1.0 µg or 10 mg (or both) per kg was used in preliminary degradation studies. Controls consisted of autoclaved hydrosoil and contained 1% $HgCl_2$. The flasks were connected to a flow-through radiorespirometer (Fig. 1) and incubated at 22°C for 28 days. Air (aerobic) or nitrogen (anaerobic) was passed through the sam-

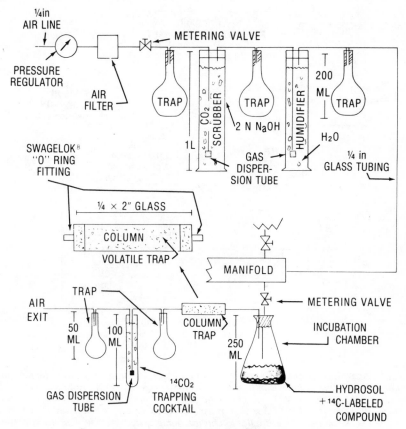

FIG. 1. Aerobic and anaerobic ¹⁴C radiorespirometer.

ples at a slow, gentle rate after first passing through an NaOH scrubber to remove CO_2. The effluent gas from the sample was passed first through a volatile absorbent trap consisting of Tenax-CG, XAD resins, or polyurethane foam. The absorbent used was adapted to the specific chemical contaminant. The gas was then finally bubbled through a $^{14}CO_2$ trapping solution of monoethanolamine-ethylene glycol. One milliliter of this solution was added to 15 ml of a methanol-fluoralloy (Beckman Instruments Inc., Fullerton, Calif.) mixture (1:1, vol/vol) in a scintillation vial and quantitated with a liquid scintillation counter. The $^{14}CO_2$ samples were taken from the trapping solution in duplicates at 1, 3, 7, 14, 21, and 28 days. Hydrosoil samples were removed for organic extraction at 3, 7, 14, and 28 days, respectively. An outline flow sheet of analysis, isolation, and identification of the degradation products is shown in Fig. 2 (19).

The HBT (or any other microbial degradation test) should present evidence not only of microbial degradation, product identification, and possible half-life of the contaminant in hydrosoil, but also of the products of simultaneous measurements of microbial biomass, chemical sorption, dissolved minerals,

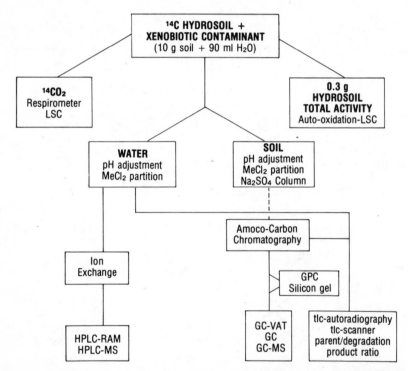

FIG. 2. *Analytical approach to the isolation and identification of organic degradation products in freshwater hydrosoil (modified from Stalling et al. [19]). Abbreviations: GC, gas chromatography; GC-VAT, gas chromatography-volatility analysis; GPC, gel permeation chromatography; HPLC-MS, high-pressure liquid chromatography–mass spectrometry; HPLC-RAM, high-pressure liquid chromatography–radioactivity monitor; tlc, thin-layer chromatography.*

biogenic and xenobiotic organics, and microbial activity. All of the factors either directly or indirectly influence the persistency of a chemical contaminant in the real world. These measurements, in my opinion, form the necessary data base for the predictive model to estimate the environmental persistency of a contaminant in aquatic ecosystems.

LITERATURE CITED

1. **Boethling, R. S., and M. Alexander.** 1979. Microbial degradation of organic compounds at trace levels. Environ. Sci. Technol. **13**:939–991.
2. **Bourquin, A. W., and P. H. Pritchard (ed.).** 1979. Proceedings of the workshop: microbial degradation of pollutants in marine environments, April 1978. U.S. Environmental Protection Agency, Gulf Breeze, Fla.
3. **Clark, R. R., E. S. K. Chian, and R. A. Griffin.** 1979. Degradation of polychlorinated biphenyls by mixed microbial cultures. Appl. Environ. Microbiol. **37**:680–685.
4. **Cochran, W. G.** 1950. Estimation of bacterial densities by means of the "Most-Probable-Number." Biometrics **6**:105–116.
5. **Giam, C. S., H. S. Chan, G. S. Neff, and E. L. Atlas.** 1978. Phthalate ester plasticizers: a new class of marine pollutant. Science **199**:419–421.
6. **Hobbie, J. E., R. J. Daley, and S. Jasper.** 1977. Use of Nuclepore filters for counting bacteria by fluorescence microscopy. Appl. Environ. Microbiol. **33**:1225–1228.
7. **Howard, P. H., J. Saxena, P. R. Durkin, and L. T. Ou.** 1975. Evaluation of available techniques for determining persistence and routes of degradation of chemical substances in the environment. Office of Toxic Substances, U.S. Environmental Protection Agency, Washington, D.C.
8. **Jannasch, H. W.** 1972. New approaches to assessment of microbial activity in polluted waters, p. 291–303. In R. Mitchell (ed.), Water pollution microbiology. Wiley-Interscience, New York.
9. **Johnson, B. T., and W. Lulves.** 1975. Biodegradation of di-n-butylphthalate and di-2-ethylhexyl phthalate in freshwater hydrosoil. J. Fish. Res. Board Can. **32**:333–339.
10. **Jones, J. G.** 1970. Studies on freshwater bacteria: effect of media composition and method on estimated bacterial populations. J. Appl. Bacteriol. **33**:679–686.
11. **Karl, D. M., and P. A. LaRock.** 1975. Adenosine triphosphate measurements in soil and marine sediments. J. Fish. Res. Board Can. **32**:599–607.
12. **Lee, G. F., and G. M. Mariani.** 1977. Evaluation of the significance of waterway sediment-associated contaminant on water quality at the dredged material disposal site, p. 196–213. In F. L. Mayer, Jr., and J. L. Hamelink (ed.), Aquatic toxicology and hazard evaluation. ASTM Special Technical Publ. 634. American Society for Testing and Materials, Philadelphia.
13. **Lenhard, G., W. R. Ross, and A. du Plooy.** 1962. A study of methods for the classification of bottom deposits of natural waters. Hydrobiologica **20**:223–240.
14. **Paul, E. A., and R. L. Johnson.** 1977. Microscopic counting and adenosine 5'-triphosphate measurement in determining microbial growth in soils. Appl. Environ. Microbiol. **34**:263–269.
15. **Pavlou, S. P., R. N. Dexter, F. L. Mayer, C. Fischer, and R. H. Haque.** 1977. Chemical-ecosystem interface, p. 21–26. In J. M. Neuhold and L. F. Ruggerio (ed.), Ecosystem processes and organic contaminants: research needs and an interdisciplinary perspective. National Science Foundation, Washington, D.C.
16. **Reid, G. K., and R. D. Wood.** 1976. Ecology of inland waters and estuaries. D. Van Nostrand Co., New York.
17. **Saeger, V. M., and E. S. Tucker.** 1973. Phthalate esters undergo ready biodegradation. Plast. Eng. **29**:45–49.
18. **Snedecor, G. W., and W. G. Cochran.** 1968. Statistical methods, 6th ed. The Iowa State University Press, Ames.
19. **Stalling, D. L., L. M. Smith, and J. D. Petty.** 1978. Approaches to comprehensive analyses of persistent halogenated environmental contaminants. ASTM Special Technical Publ. 686. American Society for Testing and Materials, Philadelphia.
20. **Wright, R. T.** 1978. Measurement and significance of specific activity in heterotrophic bacteria of natural waters. Appl. Environ. Microbiol. **36**:297–305.
21. **Zehnder, A. J. B., B. Huser, and T. D. Brock.** 1979. Measuring radioactive methane with the liquid scintillation counter. Appl. Environ. Microbiol. **37**:897–899.

Biodegradation Tests: Use and Value

P. A. GILBERT AND C. M. LEE

*Unilever Research Laboratory, Port Sunlight, Wirral, Merseyside,
L62 4XN United Kingdom*

A large number of laboratory biodegradability tests are currently used for the assessment of biodegradation potential of chemicals. To compare these existing methods, this paper attempts to clarify the methods on the basis of the information which they provide. Tests are recognized in two major categories: biodegradation potential tests and simulation tests. Tests of ready biodegradability provide convincing evidence that the material will rapidly and completely break down under environmental conditions, and tests of inherent biodegradability provide more favorable conditions for the degradation of the compound and can yield evidence of the rate of biodegradation. Under the category of simulation tests, there exist a great variety of tests designed to provide information on the rate of biodegradation under environmentally relevant conditions, and the selection of appropriate tests depends on knowledge of the expected distribution of the chemical in the environment. The need for additional research in several areas is identified in order to develop the most meaningful and predictive series of laboratory biodegradation tests.

It is important to recognize that biodegradability tests will most probably be done in the context of the hazard evaluation of a chemical, which is now accepted to be essentially the comparison of the exposure and effect concentrations (2). Biodegradation will reduce the exposure concentration, potentially increasing the safety margin attached to the particular use of a chemical, and therefore makes the task of demonstrating acceptability somewhat easier. However, it is also true that physicochemical removal, hydrolysis, and photodegradation will reduce concentrations in particular environmental compartments.

The information sought from a biodegradability test is essentially very simple. (i) Is the compound persistent? Although non-biodegradation does not necessarily mean that the compound is hazardous, where such a compound also appears to have the potential to bioconcentrate in a particular environmental compartment, lack of evidence of biodegradation indicates the need for careful consideration of this aspect. (ii) How does the observed biodegradation affect the estimated environmental concentration? (iii) Is there likely to be significant exposure to biotransformation products? There are now a large number of biodegradation tests available which provide a wide range of conditions conducive to biodegradation. The nature of the test conditions and the analytical method used will influence the degree of biodegradation observed and the type of information provided by the test.

ANALYTICAL METHODS

The analytical method used not only has a direct influence on the information obtained from the test, but also may have an indirect effect by imposing certain limitations on the test conditions.

Since it may be expensive to develop specific analytical techniques for each new chemical and its possible intermediates, many tests use the so-called nonspecific methods to follow the course of biodegradation. The most popular of these are dissolved organic carbon (DOC), oxygen uptake, and carbon dioxide evolution. These methods have the advantage of giving a measure of the complete breakdown of organic molecules and, therefore, evidence of ultimate biodegradation. They do, however, generally impose severe limitations on the test conditions; for example, at present they are most effectively used when the test compound is the sole source of carbon. The sensitivity of the method also influences the minimum test concentration which may be used. This may be important when the chemical is likely to inhibit microbial action.

In contrast, the availability of a specific analytical technique for a compound will enable biodegradation to be followed under a wide range of experimental conditions. However, there will generally be no information on the formation of intermediates and persistent residues; i.e., only evidence of primary biodegradation is obtained. The influence of choice of analytical technique on the information obtained is nicely illustrated in Fig. 1, where three very different biodegradability curves are obtained when the course of biodegradation is followed by different analytical techniques under identical experimental conditions.

The use of radiolabeled compounds, particularly if the position of the ^{14}C label is carefully considered, can provide unequivocal evidence even of ultimate biodegradability in all types of biodegradation tests. Their use seems to

BIODEGRADATION OF AN ALCOHOL ETHOXYLATE
EFFECT OF ANALYTICAL METHOD ON INFORMATION GIVEN

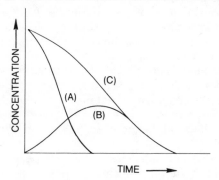

FIG. 1. *Influence of various analytical methods on the biodegradation data for an alcohol ethoxylate surfactant. (Curve A) Specific nonionic analysis: Wickbold shows primary biodegradation. (Curve B) Thin-layer chromatography: identifies polyethylene glycol residues. (Curve C) Nonspecific analysis: DOC, shows ultimate biodegradation.*

solve the problem of using a nonspecific analytical technique in the presence of high background levels of dissolved organic carbon, but there are a number of limitations (5). (i) Synthesis of labeled compounds can be complex and difficult, and the cost can be high. (ii) Complex mixtures produced in industrial processes may be difficult to reproduce at laboratory scale, and labeling is complex. (iii) Distinction between adsorption of test compound and incorporation of metabolic intermediates into new cells may be difficult without undertaking a detailed study. In practice, therefore, radiotracer techniques, although a powerful tool, are not open to routine application and are generally reserved for special research projects.

In selecting a suitable test, consideration should be given to the known properties of the chemical; for example, the respirometric methods, which depend on measurement of carbon dioxide evolution or oxygen uptake, may be the most suitable for insoluble compounds, and the closed bottle test may be used for volatile chemicals. The selection of an appropriate test method will also depend on the information required. Thus, no one test can be regarded as applicable to all chemicals.

To assist in the selection of suitable test methods, it is useful to classify them on the basis of the information which they provide (Table 1).

CLASSIFICATION OF TESTS

Several types of tests in two major categories can be recognized.

Biodegradability Potential Tests

Biodegradability potential tests give information on the susceptibility of the chemical to biodegradation by microorganisms. These tests may then be further subdivided into tests of ready biodegradability and inherent biodegradability.

(i) Test of ready biodegradability. An appropriate method is used as an initial screening test. A good result in a test of this type should provide convincing evidence that the material will rapidly and completely break down under environmental conditions, and in this case no further work is necessary. However, because of the limitations of this type of test, failure to biodegrade under these conditions does not imply that the chemical is non-biodegradable, and in this event the examination should proceed.

(ii) Test of inherent biodegradability. If under the more favorable conditions provided by these tests a positive result cannot be obtained, it must be assumed that although the compound is not proved non-biodegradable, its biodegradation may be slow or unreliable under practical environmental conditions. A negative result would therefore mean that no further work should be done on biodegradability.

A positive result indicates that the material will not persist indefinitely in the environment and, when the safety margin is large, may be all the evidence that is required. If the safety margin is small, however, and a more precise estimate of environmental concentration is required, evidence of the rate of biodegradation under relevant environmental conditions may be necessary. Such evidence is provided by simulation tests.

TABLE 1. *Classification of tests*

Test classification	Evidence provided	
	Primary biodegradation	Ultimate biodegradation
Biodegradability potential tests		
Ready biodegradability	OECD screening test (8)	Modified OECD screening test (DOC)
		Closed bottle test (O$_2$) (4)
		Closed bottle test (O$_2$) (4)
		Sturm-CO$_2$ evolution test (CO$_2$) (9)
		AFNOR T9P-302 (DOC) (7)
Inherent biodegradability	Bunch-Chambers test (1)	Zahn-Wellens test (DOC) (12)[a]
		Semicontinuous activated-sludge test with sewage feed[a]
		Modified activated-sludge test
Stimulation tests		
Biological treatment (aerobic)	SDA[b] semicontinuous activated-sludge test (10)[a]	Coupled units test (3)[a]
	OECD confirmatory test (8)[a]	
	Porous-pot test (11)[a]	
Biological treatment (anaerobic)		Anaerobic digestion test (6)
River	River die-away test (10)	
	General river elimination test (6)	
Estuary		
Sea		
Soil		

[a] For these tests it may on occasion be difficult to distinguish between biodegradation and bioelimination.
[b] SDA, U.S. Soap and Detergents Association.

Simulation Tests

Simulation tests provide information on the rate of biodegradation under environmentally relevant conditions. These may be further subdivided on the basis of the environment which they simulate. This classification allows the identification of a logical sequence of tests which may be applied in studying the biodegradability of chemicals.

The selection of appropriate simulation tests (see Table 1) clearly depends on knowledge of the expected distribution of the chemical in the environment. In cases in which the existence of a safety margin is still in doubt, it may be necessary to seek further confirmation by undertaking field trials and appropriate monitoring.

This approach seems reasonable for the examination of a large number of different types of chemicals, but it raises a number of fundamental questions.

What are the characteristics of a test of ready biodegradability?

A test of ready biodegradability has to be stringent and failsafe in the sense that all chemicals identified as readily biodegradable do rapidly and completely break down under environmental conditions. Recognition of tests of ready biodegradation is empirically based on limited knowledge and experience. Suitable tests seem to have the following characteristics. (i) They are based on a nonspecific method of analysis; i.e., they can be applied to a wide range of organic chemicals without the need for the development of specific analytical procedures. They give an indication of the extent of ultimate biodegradation. (ii) They involve exposure of a relatively low concentration of test chemical to a small number of bacteria in the absence of other organic nutrients, i.e., no opportunity for cometabolism. (iii) The test period is normally 3 to 4 weeks, and at best only a limited opportunity for preacclimatization of the inoculum is provided.

What level of biodegradation is indicative of ready biodegradability?

Experience has shown that 100% loss of DOC is rarely achieved in practice since water-soluble products may be synthesized during the growth of the bacteria that do not biodegrade in the time scale of the test. Even in the case of glucose, for example, the observed loss of organic carbon lies in the range of 85 to 95%. In the respirometric tests an even lower result should be expected since any part of the chemical that is used for the synthesis of new cell materials will not be measured as biodegradation in these tests. Figure 2 shows typical results of a CO_2 evolution test in which aniline, glucose, and 4-nitrophenol were examined, illustrating the expected shortfall in removal of these soft standards.

The setting of a unified "pass" level is further complicated by variation in the test conditions provided in different tests, the duration of the tests, and the opportunity provided for acclimatization. Such a level can only be set on an arbitrary basis, but it must reflect a balance between setting the level high enough to ensure that compounds producing persistent residues do not pass and not so high as to "fail" many compounds that are readily biodegradable in practice. On the basis of the limited experience available, it is suggested that ready biodegradability could be assumed for chemicals giving greater than 70% biodegradation in a DOC test or greater than 60% biodegradation in a respirometric test. It should be emphasized, however, that these values should not be viewed in isolation. For example, the shape of the biodegradation curve may also be considered. This is exemplified in the case of the nonionic surfactant (Fig. 2) which is still showing evidence of continuing biodegradation at the end of the test. The structure of the chemical and the possibility of the formation of recalcitrant structures should also be considered. In the case of a pure compound containing 20 carbon atoms and assuming 60% of the molecule has been converted to CO_2 and 10% to cell metabolites, the 20% unaccounted for could only contain a theoretical maximum of 6 carbon atoms.

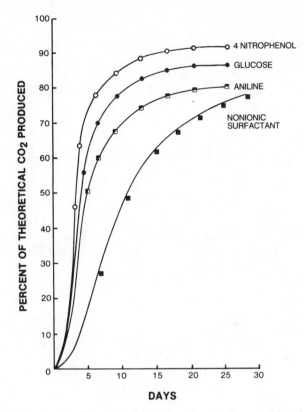

FIG. 2. *Comparative data for several organic compounds tested for ultimate biodegradability expressed as CO_2 production.*

The question then posed is, Is a recalcitrant structure feasible which contains only 6 carbon atoms?

How do the results of tests of ready biodegradability relate to behavior in the real environment?

Practical experience has shown that chemicals which exhibit 60 to 70% biodegradation in these tests would be better than 90% removed during biological treatment of sewage. The results summarized in Table 2 illustrate this point. In the case of diethylene glycol, for example, although there is some evidence of biodegradation in these simple tests, the results are variable and generally well below the suggested pass level. However, under activated-sludge conditions almost complete biodegradation is observed. Diethylene glycol is an example of a material to which bacteria seem to acclimatize easily but which is slow to degrade under the test conditions, whereas with oxydisuccinate no evidence of acclimatization or biodegradation was observed in these simple tests. In this latter case, however, the compound can be shown to be biodegradable, and good removal was observed in a sewage treatment plant, at least under favorable operating conditions.

TABLE 2. Biodegradation of a number of chemicals in the various test classes

Chemical	Structure	Tests of ready biodegradability	Tests of inherent biodegradability	Simulation tests	Comments
Carboxymethyloxysuccinate	CH_2-CO_2H $-O-$ $-CH-CO_2H$ $-CH_2-CO_2H$	100% (a)	100% (b)	100% (c)	(a) River die-away test (b) SCAS test[a] (c) LASP method[b]

Nitrilotriacetic acid	CH_2-CO_2H
	$-N-CH_2-CO_2H$
	$-CH_2-CO_2H$
		1. 0 (a)
		2. 100%
			100% (b)	1. 60% (c)
				2. 100%
					(a) River die-away test
					   1. Unacclimatized bacteria
					   2. Acclimatized bacteria
					(b) SCAS test
					(c) 1. 10°C recovered to 100% at 20°C
					   2. 20°C

Diethylene glycol	CH_2OH
	$-CH_2$
	$-O-$
	$-CH_2$
	$-CH_2-OH$
		16–63% (a)	100% (b)	1. 88% (c)
				2. 52%
					(a) Variable depending on test method
					(b) Zahn-Wellens test
					(c) Husmann test
					   1. Domestic sewage
					   2. Artificial feed, OECD

Compound	Structure				Test methods
Oxydisuccinate	CH₂—CO₂H CH—CO₂H —O— CH—CO₂H CH₂—CO₂H	0 (a)	~90% (b)	1. 75% (c) 2. 90%	(a) River die-away test (b) SCAS test after 4 weeks (c) LASP method with 1. 6-h detention time 2. 12-h detention time
Cyclopentane tetracarboxylate		0–30% (a)	~85% (a)		(a) Depends on the stereoisomers used (b) SCAS test after 8 weeks
Ethylene glycol dimalonate	CO₂H CH—O—CH₂—CH₂—O—CH CO₂H CO₂H CO₂H	0 (a)	0 (b)		(a) River die-away test (b) SCAS test

^a SCAS test, Semicontinuous activated-sludge test.   ^b LASP, Laboratory-scale activated sludge plant.

There is also evidence that these simple tests of ready biodegradability offer less favorable conditions for biodegradation than would exist in a river. Figure 3 shows the biodegradation of an alcohol ethoxylate in the Organization for Economic Cooperation and Development (OECD) screen test and in a river die-away test using real river water. Primary biodegradation, although somewhat more rapid in the river water, still occurs to the extent of ~90% in the OECD screen test within a few days. However, when the formation and further biodegradation of the polyethylene glycols in the tests are compared, it is found that, although the maximum amount of polyethylene glycol formed is higher in the river water, the subsequent biodegradation occurs more rapidly than in the artificial test medium.

What are the characteristics of a test of inherent biodegradability?

Tests of inherent biodegradability should provide the most favorable conditions for acclimatization and biodegradability. (i) The compound should be exposed to as wide a range of bacterial species as possible over a prolonged period of time. (ii) The ratio of biomass to compound should be relatively large. (iii) Alternative sources of organic nutrients should be available to maintain the viability of the bacterial inoculum and to provide opportunities for cometabolism.

The most suitable test currently available seems to be the modified semi-continuous activated-sludge test. In this method settled sewage and the test compound are aerated for 24 h. The bacterial flocs are then settled, and the DOC of the supernatant is determined and compared with that of a control experiment. More settled sewage and test compound are then added, and aeration is restarted. Each 24-h period might therefore be regarded as a new test with some reinoculation, but bacteria which have been exposed to the

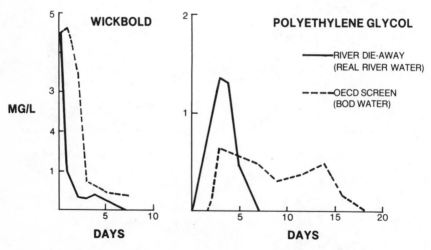

FIG. 3. *Effect of varying test conditions and different inocula on the biodegradation rate of an alcohol ethoxylate surfactant. BOD, biochemical oxygen demand.*

test compound are carried over each time. In each 24-h period conditions change significantly, so that after ~6 h nutrient concentrations will be low and the bacteria are under considerable pressure to adapt to the test compound. Thus, the test provides good conditions for acclimatization but provides only 24 h for biodegradation.

Biodegradation may be followed by measuring loss of total organic carbon or DOC, but this method requires that a relatively high level of test compound be added due to the high background levels of organic carbon. If an appropriate specific analytical technique is available, more realistic levels of test compound may be used, but this may limit the information gained from the test to primary biodegradation, whereas organic carbon analysis may detect ultimate biodegradation.

The results shown in Table 2 for oxydisuccinate and cyclopentane tetracarboxylate clearly demonstrate the power of this test. However, there are only a limited number of tests of this type available, and further development would be useful.

What are the characteristics of simulation tests?

The characteristics of simulation tests will vary with the environmental compartment it is desired to simulate, but for all tests of this type evidence showing their relationship to the environment should be available. Here, discussion will be limited to those methods used to simulate the activated-sludge treatment of sewage.

The method used in Europe to determine the biodegradability of surfactants is the Husmann test (4), which utilizes an artificial nutrient feed inoculated with sewage organisms from secondary effluent. The level of test compound remaining in the effluent is followed over several weeks to determine the average removal. This is a satisfactory method for controlling the biodegradation of a group of closely related chemicals, such as the surfactants, for which there is detailed knowledge of the practical behavior of some members of the group. An interesting variant of the method has recently been developed by Fischer and Gerike (3), who have adapted it to allow DOC analysis to be used to follow biodegradation. In this method half the activated sludge is interchanged daily between two units to reduce the variability between the control and experimental units. The Husmann approach does, however, suffer from a number of problems which affect its application to chemicals in general:

1. The nature of the activated sludge and its settling characteristics are very different from that formed with a real sewage feed, and there is some evidence that the real sewage feed gives more favorable results since it has been observed, for example, that the average removal of diethylene glycol in a 5-week period averaged 88% in domestic sewage but only 52% with the OECD feed.

2. The residence time in the Husmann unit is relatively short (3 h) compared to normal practice (3 to 6 h in Germany, 6 to 12 h in the United Kingdom). In some cases this may have a profound effect on the level of removal observed. Fischer and Gerike (3), for example, observed

70% mineralization of linear alkylbenzene sulfonate with a 3-h residence time but 94% with a 6-h residence time.
3. Although a settler is incorporated into the test apparatus, the residence time is again unrealistic (~2 to 3 min compared to 3 h in practice).
4. Most importantly, variations in loading, sludge age, and temperature cannot be simulated in a Husmann unit.

For these reasons the porous pot (11) with real settled sewage is preferred as a simulation test. With this system the time taken for and the stability of acclimatization to the compound and the removal rate of the compound can be followed under a typical range of operational conditions. The results of these tests provide valuable evidence which enables considerable refinement of the calculation of exposure levels in the environment.

SUMMARY

Although a large number of biodegradability tests have been developed in the literature, it is possible to simplify the picture by classifying them on the basis of the information which they provide. Such a classification then enables a logical test sequence to be identified and used in studying the biodegradability of chemicals. The approach is, however, empirical, and so far it is based on relatively limited experience. The following questions therefore remain.

1. Ring tests of biodegradability test methods usually demonstrate considerable interlaboratory variability in the results obtained. More work is therefore needed to identify the reasons for this and to make proposals to reduce it, for example, by standardization of test media and of appropriate inoculation procedures.
2. What are the most favorable conditions which should be used in determining the inherent biodegradability of chemicals?
3. What are the most relevant and cost-effective methods of simulating the various environmental compartments and how can information on the rate of biodegradability obtained in the laboratory be used to predict behavior in the environment?

LITERATURE CITED

1. **Bunch, R. L., and C. W. Chambers.** 1967. A biodegradability test for organic compounds. J. Water Pollut. Control Fed. **39**:181–187.
2. **Cairns, J., Jr., K. L. Dickson, and A. W. Maki.** 1978. Summary and conclusions, p. 191–197. *In* J. Cairns, K. L. Dickson, and A. W. Maki (ed.), Estimating the hazard of chemical substances to aquatic life. ASTM Special Technical Publ. 657. American Society for Testing and Materials, Philadelphia.
3. **Fischer, W. K., and P. Gerike.** 1975. Biodegradability determinations via unspecific analysis (chemical oxygen demand, dissolved organic carbon) in coupled units of the OECD confirmatory test. Water Res. **9**:1131–1135.
4. **Fischer, W. K., P. Gerike, and R. D. Smid.** 1974. Combination of methods for successive testing and evaluation of the biodegradability of synthetic substances as organic complexing agents by means of generally valid nonspecific parameters (BOD, CO_2, COD, TOC). Wasser Abwasser **7**:99–118.

5. **Gilbert, P. A., and G. K. Watson.** 1977. Biodegradability and its relevance to environmental acceptability. Tenside **14:**171–177.
6. **Janicke, W., and G. Hilge.** 1977. Bestimmung des eliminationsgrades wassergefahrdender Stoffe. Wasser Abwasser **10:**4–9.
7. **Laboureur, P.** 1977. Biodegradabilité. Methods d'evaluation et methode de determination, T91L-10. L'Association Francaise de Normalisation, Paris.
8. **OECD.** 1971. Pollution by detergents. Determination of the biodegradability of anionic surface active agents. Organization for Economic Cooperation and Development, Paris.
9. **Sturm, R. N.** 1973. Biodegradability of nonionic surfactants: screening test for predicting rate and ultimate biodegradation. J. Am. Oil Chem. Soc. **50:**159–167.
10. **U.S. Soap and Detergents Association.** 1965. A procedure and standards for the determination of the biodegradability of alkyl benzene sulphonate and linear alkylate sulphonate. J. Am. Oil Chem. Soc. **42:**986–993.
11. **Water Research Centre.** 1978. WRC porous-pot method for assessing biodegradability. Technical Rep. no. TR70. Water Research Centre, Stevenage, Hertfordshire, England.
12. **Zahn, R., and H. Wellens.** 1974. A simple method of determining the biodegradability of products and contaminants in wastewater. Chem. Ztg. **98:**228–232.

Synopsis of Discussion Session: Biodegradation Methodology

W. E. GLEDHILL, B. T. JOHNSON, M. KITANO, AND C. M. LEE

Previous conferences in Pensacola, Fla. (5), Pellston, Mich. (1), and Waterville Valley, N.H. (2), have dealt with various aspects of the hazard evaluation process, which is defined as a determination of the fate and effects of chemicals in the environment. Biodegradation was identified as one of the major processes which determine the fate of a chemical in the environment, and it was the assignment of this task group to evaluate currently available methods, develop evaluation criteria, and rank existing methods.

The objective of biodegradation testing is to assess the extent to which microflora can alter a chemical or mixture of chemicals. The importance of biodegradation is that it is often the key process for reducing exposure concentrations in the environment, although it need not necessarily be considered a prerequisite for environmental acceptability.

BIODEGRADATION TEST METHODS

Over the past two decades numerous biodegradation tests have been developed, thus allowing for considerable flexibility in the selection of appropriate tests to take account of the structure, properties, and expected use of the chemical under review. It is important to remember the practical constraints on the examination of large numbers of new chemicals and therefore to do the tests in a logical sequence which minimizes cost and effort.

A sequence has been proposed (Table 1) in which the first step is to determine whether the chemical is susceptible to breakdown by microorganisms. Tests in this group can be further divided into those offering stringent conditions and those offering very favorable conditions for degradation ("ready" and "inherent" biodegradation, respectively).

Existing stringent methods are characterized by the use of simple, synthetic media in which the test chemical is effectively the sole carbon source (1 to 20 ppm) and a low biomass (e.g., 10^2 to 10^5 cells per ml) is used. The inoculum should be obtained from the most appropriate source depending on the expected use of the chemical, e.g., river water, soil, domestic sewage. To maintain stringent conditions, no preacclimation of the inoculum is made.

The analytical method has a significant influence on the perceived biodegradation so that methods which indicate ultimate biodegradability are preferred since these are more persuasive in a hazard evaluation context. Experience so far suggests that chemicals which exhibit substantial loss of dissolved organic carbon (DOC), production of theoretical CO_2, or uptake of oxygen in these tests achieve a very high level of biodegradability in the environment, and further testing may be unnecessary.

Under these stringent conditions, a positive result may, therefore, be interpreted as indicating that the chemical is susceptible to ready biodegradation. On the other hand, a negative result need not mean that the chemical

46

TABLE 1. *Biodegradation test procedures*

Tier	Test
I	*Biodegradation "potential"*
	"Ready" biodegradation (low initial biomass)
	Synthetic media
	Complex media
	"Inherent" biodegradation (high biomass)
	Activated-sludge die-away
	Aquatic sediment die-away
	Soil die-away
	Anaerobic screening test
	Chemostats
	Pure cultures
II	*Simulated natural environments*
	Continuous-flow biological treatment systems
	Anaerobic biological treatment systems
	Microcosms (natural water + sediments)
	Freshwater
	Marine
	River water
	Soil systems
III	*Field studies*
	Sewage treatment plants
	Natural waters
	Soils

is non-biodegradable, but only that more favorable conditions may be required.

In the latter case, the chemical may be submitted to a test in which all possible favorable conditions are provided. Existing tests of this type are characterized by particularly high biomass (10^7 to 10^9 cells per ml) and the presence of alternative sources of carbon, giving an opportunity for cometabolism. These tests are usually run for a prolonged period (up to 3 months) to give an opportunity for acclimation. This length of time is not regarded as an environmentally significant issue provided that the subsequent rate of biodegradation is acceptable and acclimation of the receiving system is maintained. Specific analytical methods and radiochemical techniques are often favored in these tests since, at present, it is difficult to apply nonspecific methods such as DOC or CO_2 measurements to chemicals at environmentally realistic levels as a consequence of high levels of background carbon. These tests seek to show only whether or not the chemical is inherently susceptible to breakdown by microorganisms, and, therefore, no specific "pass" level is set. Failure to degrade may not mean the chemical is non-biodegradable, but only that the degradation rate may be too slow or unreliable. A positive result indicates that the chemical will not persist indefinitely in the environment, but evidence for the rate of biodegradation or removal of the chemical in nature may be needed. To obtain this, chemicals are submitted to an appropriate simulation test (tier II) depending on their expected use pattern.

The objective of this second major group of tests is to simulate particular compartments of the receiving environment (e.g., sewage treatment, soil, sediment, river water). There is considerable need for the further development and validation of these tests, which are difficult to standardize and which are currently used primarily as research tools.

Finally, in some special cases it may be necessary to proceed to actual field conditions (tier III) to provide satisfactory evidence for the adequate removal of the chemical. However, it should be recognized that this activity mainly provides a check of the environmental exposure concentration and that any perceived reduction may be a result of physiochemical (e.g., adsorption) removal as well as biodegradation.

APPLICABILITY OF THE TESTING SCHEME

Examples of three types of organic chemicals will be used to demonstrate the utility of the tests in Table 1: a detergent product such as a surfactant, an agricultural chemical such as a herbicide, and an industrial chemical such as a plasticizer. For each of these chemicals, initial screening for ready biodegradation would be carried out in a synthetic medium. The medium would receive inoculum from the likely receiving environment for the chemical. For the above three compounds, secondary sewage effluent, soil microorganisms, and river water, respectively, might serve as suitable inocula. Nonspecific analytical techniques would be used to assess O_2 uptake, CO_2 evolution, or DOC disappearance. Chemicals displaying in excess of 60% theoretical O_2 uptake or CO_2 evolution would be regarded as readily biodegradable. Little additional work on these materials would be warranted.

Slowly degradable chemicals would be subjected to an inherent biodegradation test. For the detergent product, an activated sludge die-away could be chosen, and a soil die-away and aquatic sediment die-away could be chosen for the agricultural and industrial chemicals, respectively. All materials would be screened for anaerobic biodegradability. Materials identified as non-biodegradable in these tests would not be tested further for biodegradability. Materials showing degradation in the high-biomass test systems (inherent biodegradability) would be subjected to tier II testing in simulated natural environments. Tests chosen would be based on the anticipated receiving environment for the chemical. A detergent product might be examined in aerobic and anaerobic sewage treatment systems, in rivers, and in a freshwater microcosm. The agricultural chemical would be subjected to more detailed studies in soil and river water. The industrial chemical might be examined in a freshwater microcosm and river water. Biodegradation studies would be terminated for chemicals showing "inadequate" biodegradation in these simulated systems.

KEY ISSUES

Discussions focused on a number of particularly important questions.

What analytical data are needed to support biodegradation tests?

The significance of the analytical method in determining the perceived level of biodegradation has already been noted. In relatively simple test systems (low biomass, chemically defined media), such as tests to measure oxygen uptake or CO_2 evolution, nonspecific analytical techniques can be used to establish the extent of mineralization. In complex systems (high biomass, complex media), such as activated sludge or sediment-water mixtures, specific analytical techniques are usually required for extraction and measurement of the test chemical. Without the use of radiolabled substrates, assessment of biodegradability in high biomass must rely on extraction procedures and specific chemical analyses (primary biodegradation). More sensitive O_2 uptake and CO_2 evolution methods are under development which may eventually allow the measurement of ultimate biodegradation by use of nonspecific techniques.

What current or evolving technology applies to biodegradation of insoluble materials, volatile materials, anaerobic systems, sediments, and marine and estuarine systems?

(i) **Insoluble chemicals.** With the exception of those methods based on DOC analysis, most methods applicable to water-soluble chemicals can be used to assess the biodegradability of chemicals having low aqueous solubilities.

Various methods have been used to add slightly soluble chemicals to test systems, but it was concluded that the best approach is to add the pure chemical to the system. Sonication can be used to disperse or emulsify the chemical in the test medium prior to inoculum addition, or if sonication is not effective, adsorption of the test chemical to an inert carrier, such as sand, can be used. Preferably, solvent carriers should be avoided.

Toxicity problems, i.e., inhibitory effects of the test chemical on the inoculum, may be greater with chemicals having low aqueous solubility. Insoluble chemicals, such as organochlorine pesticides (e.g., Kepone), may accumulate at high concentrations on lipophilic cell surfaces. Incremental substrate addition or use of lower concentrations of radiolabeled chemicals may circumvent these problems.

(ii) **Volatile chemicals.** The biodegradability of volatile chemicals can be measured in closed systems such as certain O_2 uptake and CO_2 evolution procedures. In complex systems, specific analytical methods and detailed material balances must be critically examined.

(iii) **Anaerobic test procedures.** Anaerobic biodegradation test procedures in systems simulating a wastewater treatment plant anaerobic digester are well documented in the literature. These procedures are useful in measuring the extent of mineralization (CO_2 + CH_4 production) of a test compound. The limits of treatability may also be determined as the concentration inhibiting important stages of the treatment processes is approached. The procedure recently described by Young and co-workers (3, 4) offers a simple method for screening the anaerobic biodegradability of chemicals. For certain compounds, such as those containing nitro groups, consideration should be given to a combination of anaerobic and aerobic processes to assess biodegradability.

A general limitation of anaerobic biodegradation procedures has been the need for relatively high concentrations (>100 ppm) of test chemical. Also, in many studies there has been no assurance that anaerobic conditions were maintained throughout the test period. Consideration should be given to the use of oxidation-reduction indicators, such as methylene blue or resazurin, to confirm that anaerobic conditions are maintained. Inocula for anaerobic studies should be from anaerobic systems.

(iv) Sediments. This conference has addressed studies of chemicals in sediment-water systems, and, in general, a specific analytical method or radio-labeled material is required to follow the degradation of chemicals in these systems. For certain chemicals, such as new plasticizers, dielectric fluids, etc., consideration should be given to the use of these tests early in the screening phase (tier I) of a multitiered testing program.

(v) Marine-estuarine systems. Saltwater systems are important natural environmental compartments. In general, methods useful for assessing the biodegradability of chemicals in freshwater systems are also applicable to saltwater systems. When using salt water in biodegradation studies, it should be recognized that these systems are often deficient in nitrogen and phosphate. Supplementation of test systems with these nutrients may be necessary, but may affect the environmental relevance of the particular test.

What is the role of microcosms or other multicompartment model systems in biodegradation studies?

Other discussion groups will address this question in more detail. Microcosms are used to simulate an environmental compartment in the laboratory and potentially can more realistically reflect biodegradation or removal rates in the actual environment.

Perhaps the most studied and closest approximation of a "real world" situation is the simulation of sewage treatment systems. We are fairly capable of predicting "real world" behavior of sewage systems from laboratory studies. As our experience increases, microcosms simulating natural water systems may become more useful in predicting the fate of chemicals in nature.

There have been many attempts to simulate natural compartments, but few, if any, attempts have been made to characterize systems in terms of limnological or geological parameters. From a practical viewpoint, microcosms may be the closest approximation to natural systems that can be studied in the laboratory. They should be essential tools in the decision-making process for development of new products.

When do breakdown products become important and what methods should be used to test biodegradation intermediates?

Studies of biodegradation intermediates become important if these materials are not part of natural biodegradation cycles or are more persistent, toxic, or bioaccumulative than the parent chemical. Indications that biodegradation intermediates may need further consideration are (i) complex chemical structures which could require a complex degradation pathway, (ii) the presence of branched structures, (iii) halogens or polycyclic rings, (iv) intermediate results in tier I tests, (v) more than one stage of breakdown, and (vi) disparity between results of primary and ultimate biodegradation tests. When significant intermediates are identified, they should be subjected to the same testing regime as the parent compound. It may be desirable to establish whether the intermediates are completely refractory or satisfactorily degradable under more natural conditions (microcosms).

By-products in large-volume commercial chemicals should also receive consideration when present in signficant quantities (>1%).

Can compounds which are used as sole source of carbon by pure or mixed cultures be considered biodegradable and therefore environmentally acceptable?

In general, experience suggests that chemicals used as the sole carbon and energy source by pure or mixed cultures will be biodegradable in the environment, but there could be exceptions. As an example, with some chemicals (polychlorinated biphenyls, for example), detailed enrichment procedures may select slow-growing, nutritionally fastidious species capable of utilizing a test substrate as the sole carbon source. However, as a result of input rate and slow biodegradability, these materials may behave poorly in the natural environment. Moreover, chemicals with these properties would be detected in the tier I tests. Any evidence of biodegradability makes the task of demonstrating environmental acceptability of a chemical easier but cannot be used as the sole determinant of environmental acceptability.

How does cometabolism influence biodegradability determinations?

In simple tests where the chemical is the sole source of carbon, cometabolism cannot play a part in biodegradation. In tests with complex media and in simulation tests, cometabolism may occur. The implications of the need for cometabolic processes to break down a chemical in the open environment are not understood.

MAJOR CONCLUSIONS

1. Biodegradation is often the key environmental process for removal of synthetic organic chemicals, and adequate methodology has been developed to assess this property. Methods which give information on ultimate biodegradation (mineralization) are more persuasive in a hazard assessment context.
2. Chemicals which substantially degrade in synthetic media inoculated with low biomass from a natural environment can be considered readily biodegradable.
3. A positive result in any biodegradation test indicates that a chemical may not persist indefinitely in the environment. The exposure concentration is determined by the rate at which a chemical is degraded and the rate at which it is added to the system.
4. Current biodegradation test methodology can detect production of intermediates or residues which require further examination.
5. Biodegradation is *not* a prerequisite for environmental acceptability of a chemical. However, a less detailed environmental assessment program is required for a readily biodegradable chemical.
6. The need for acclimation is not a significant issue provided that the subsequent rate of biodegradation is acceptable and acclimation of the receiving stream is maintained.

RECOMMENDATIONS

1. Tests for ready biodegradability should have good inter- and intralaboratory reproducibility. The inoculum may be the single most important variable in these tests. Initial screening studies should not use "special species" of microorganisms. Natural mixed microbial populations from soil, sewage systems, sediments, and natural waters should be used.
2. The Pensacola meeting (5) concluded that excessive emphasis has been placed on standardization and uniformity of methods. In general, the participants agree, since it is difficult to achieve a standard microbial inoculum. Lack of standard inocula could be compensated for by use of standard reference chemical. Consideration should be given to the development of an adequate set of standards.
3. Ready biodegradability tests should use nonspecific analysis, such as O_2 uptake, DOC disappearance, or CO_2 evolution, to estimate the extent of mineralization. More sensitive automatic instruments should be developed so that more realistic chemical concentrations can be studied. Automatic measurement of anaerobic biodegradability ($CO_2 + CH_4$) is also desirable.
4. Confirmation of laboratory studies via periodic field monitoring is required. This will serve as a check on the rationale for use of certain biodegradation procedures.
5. Further effort is needed to design laboratory simulations of natural compartments. Consideration of geochemical and limnological properties may be of value in interpreting the laboratory data in a natural community.

LITERATURE CITED

1. **Cairns, J., K. L. Dickson, and A. W. Maki (ed.).** 1978. Estimating the hazard of chemical substances to aquatic life. ASTM Special Technical Publ. 657. American Society for Testing and Materials, Philadelphia.
2. **Dickson, K. L., A. W. Maki, and J. Cairns (ed.).** 1979 Analyzing the hazard evaluation process. American Fisheries Society, Bethesda, Md.
3. **Healy, J. B., Jr., and L. Y. Young.** 1979. Anaerobic biodegradation of eleven aromatic compounds to methane. Appl. Environ. Microbiol. **38**:84–89.
4. **Owen, W. F., D. C. Stuckey, J. B. Healy, Jr., L. Y. Young, and P. L. McCarty.** 1979. Bioassay for monitoring biochemical methane potential and anaerobic toxicity. Water Res. **13**:485–492.
5. **U.S. Environmental Protection Agency.** 1979. Microbial degradation of pollutants in marine environments. Proceedings of the Workshop, Pensacola, Fla. EPA 600/9-79-012. Environmental Protection Agency, Washington, D.C.

Chapter 4. BIOAVAILABILITY AND THE MATERIAL BALANCE APPROACH TO ESTIMATING AQUATIC EXPOSURE CONCENTRATIONS

Material Balance Approach to Estimating Aquatic Exposure Concentrations

DEAN R. BRANSON AND GARY E. BLAU

Dow Chemical Company U.S.A., Midland, Michigan 48640

Concentrations of chemicals in water lost from process plants can be estimated with the aid of a material balance (D. M. Himmelblau, *Basic Principles and Calculations in Chemical Engineering,* 1972). Application of this engineering practice to the field of environmental science illustrates the benefits of interdisciplinary joint ventures. During the workshop we presented a material balance case study involving pentachlorophenol. The purpose of the study was to estimate the difference, if any, in environmental hazard between distilled and undistilled pentachlorophenol when used to treat wood via the pressure treatment process. Distillation reduces the level of pentachlorophenol impurities, including chlorinated dioxins, at least tenfold. This brief summary will focus primarily on the key elements of the material balance approach.

MATHEMATICAL DESCRIPTION OF A PROCESS

A process is a flow of materials through one or more operations to achieve a desired result. In designing and controlling a process, the engineer can optimize the overall process performance according to any combination of several criteria, including conversion, efficiency, dollars, time, and waste generated. Each operation in the process can be described as a set of equations, i.e., a mathematical model. These equations describe the output from each operation, given the input and certain operating conditions such as temperature. The environmental scientist can extract from this information an estimate of the amount of material lost from the process in the form of air emissions, wastewater discharges, and other wastes. It follows that calculating an aquatic exposure concentration can be accomplished by linking the amount of material lost from a process to any of several environmental fate models.

The first step in performing a material balance on a process is to prepare a flow diagram. In the case of the pentachlorophenol (penta) study, the process flow diagram contained three major engineering operations: a mixing tank for penta and the oil solvent, a treating cylinder, and an oil-water separating device.

The next step is to apply the first law of thermodynamics, i.e., the law of conservation of mass, to each chemical present in each operation. This set of equations describing the process must also obey certain constitutive relationships. For example, in the penta wood-treating process, it is important to characterize the enrichment of penta impurities in the oil solvent and sludge.

53

Phase-constitutive relationships are generally fundamental principles of physical chemistry, e.g., phase equilibrium. If actual data are not available, then the necessary values can, and must, be estimated. This was the case for all of the penta impurities. Vapor pressures at 70°C, the typical temperature in a treating cylinder, were estimated for each of the components, using Lyderson group contribution methods with the Riedel-Plank-Miller vapor pressure correlations (3). The estimated oil-water partition coefficients for the penta impurities were based on group contribution theory and calculated by the UNIFAC method. This method uses a data base that contains interaction energies from published vapor-liquid equilibrium data for about 2,500 compounds (1).

EXAMPLES OF RESULTS FROM THE PENTA CASE STUDY

Applying the material balance approach, we discovered that an average-sized plant engaged in treating wood with penta under pressure can be described as follows: 1,800 ft^3 of wood treated with 900 pounds of penta per day, 0.5 pound of penta per ft^3 of wood, 1.5 gallons of wastewater per ft^3 of wood, 4.9 $\times$ 10^{-3} ft^3 of waste sludge per ft^3 of wood. Once the equations for such a plant have been written down, they can be solved to calculate the amount of the individual components leaving each operation of the process.

For the wood-treating process, the penta and impurities leaving the plant were in the wood (>99.5%), in the wastewater (<0.1%), or in the waste sludge (<0.4%). The amount of octachlorodioxin leaving this typical plant in 2,700 gallons of wastewater was 5 $\times$ 10^{-11} pounds each day. With an oil-water partition coefficient of 10^7 for octochlorodioxin, this translates into an aquatic exposure concentration of 9 $\times$ 10^{-2} ng/liter (ppt). This example is for 900 pounds of undistilled penta with 1,000 ppm of octachlorodioxin.

When any estimation procedure is used to determine constitutive relationships, the concentration predicted by the model should be validated with several measured values, if possible.

Examples of other early-stage environmental questions which can be addressed in an analytical framework with the aid of a material balance include the following. (i) What are the expected benefits of specific pollution control equipment? (ii) What are the critical chemical impurities of a mixture? (iii) Will product A or B cause greater exposure of humans and the environment? (iv) What additional data, if any, are necessary to assess the hazard of the chemical(s) in a specific aquatic environment?

CONCLUSIONS

The conclusions from the penta case study were that distilled penta would not result in a significant difference in environmental hazard compared with undistilled penta. The penta impurities have a high affinity for the oil solvent, which is separated from the water before discharge. The resulting concentrations of penta impurities in receiving water are predictably so low as to be biologically insignificant, especially compared with the ingredient controlling toxicity—pentachlorophenol itself.

LITERATURE CITED

1. **Fredenslund, A., R. L. Jones, and J. M. Prausnitz.** 1975. Group contribution estimation of activity coefficients in nonideal liquid mixtures. AIChE **21**:1086–1099.
2. **Himmelblau, D. M.** 1972. Basic principles and calculations in chemical engineering, 3rd ed. Prentice-Hall, Inc., Englewood Cliffs, N.J.
3. **Reid, R. C., J. M. Prausnitz, and T. K. Sherwood.** 1977. The properties of gases and liquids. McGraw-Hill Book Co., New York.

Bioavailability of Chemicals in Aquatic Environments

JERRY HAMELINK

Lilly Research Laboratories, Division of Eli Lilly and Company, Greenfield, Indiana 46140

The bioavailability of chemicals in aquatic environments will depend on a number of factors which interact to control the external availability of a chemical to an aquatic organism. This paper addresses those factors influencing the dissolved fraction of chemicals in water and the ways in which a chemical associated with sediments becomes available to aquatic species. Bioavailability is considered as an exchange process whereby a xenobiotic must pass between a boundary layer separating an abiotic medium and the organism. It is shown that the rate of passage through the boundary layer can be approximated by a modification of the evaporative loss equation.

Bioavailability, for purposes of this discussion, will be limited to consideration of what factors determine the external availability of a chemical to an aquatic organism, as opposed to the more classic pharmacological consideration of internal bioavailability after injection or ingestion. This limited definition is intended to focus discussion on what factors determine the spacial, temporal, and concentration gradients between aquatic organisms and chemicals.

Interactions between solids and water often influence the bioavailability of chemicals to organisms in aquatic environments. Sediments frequently transport chemicals out of the water column, such that bottom sediments generally serve as a "sink" for xenobiotics. Yet some particularly troublesome environmental problems have been traced to situations where sediments have served as a source of xenobiotics. Hence bioavailability, as it relates to hazard evaluation, has frequently focused on the question of when suspended solids and sediments serve as "sources" or "sinks" of trace chemicals in aquatic environments.

The accumulation of chemicals by free-living aquatic invertebrates and fish is assumed to be controlled by the truly dissolved concentration of chemicals in the water (7, 12). The recognized difficulty with this assumption is that chemicals associated with sediments may be taken up by a variety of organisms. In addition, the concentration and persistence of chemicals in water depend upon many factors (1), including sediment-water interchange. Thus, this paper will first address those factors that influence the truly dissolved fraction of chemicals in water and then the question of how a chemical associated with sediments becomes available to a benthic organism, by ingestion or by direct uptake from the interstitial water.

RATIONALE

Consider bioavailability to be an exchange process whereby a xenobiotic must pass between a boundary layer separating an abiotic medium (A) and a biotic object (B):

Then the question becomes, what controls passage into, through, and out of the boundary layer?

The rate of passage through the boundary layer can be approximated by a modification of the evaporative loss rate equation (14). This equation assumes that the flux of chemical through the boundary layer is proportional to the concentration gradient across the layer (i.e., Fick's law of diffusion applies). This is reasonable once the chemical enters the boundary layer, which is assumed to be composed of water for purposes of this discussion. Unfortunately, I doubt that bioavailability is generally controlled by the rate of passage through the boundary layer or layers. Rather, the transfer rate is probably controlled by the rate at which chemicals enter and leave the boundary layer. Hence, quantification of bioavailability may largely depend on measurement of the transfer rates into and out of the boundary layer.

Perhaps an easy way to quantify the transfer rate into the boundary layer can be devised by incorporating some of the principles developed by soil scientists for predicting leaching and mobility of chemicals in soils. This behavior can be estimated by soil thin-layer chromatography (TLC) R_f measurements (5), the omega index (9), and in several other ways, including the use of free energy concepts (4). Chemicals which can be readily moved through soil by water must have either little affinity for the soil or a greater affinity for the water. Thus, those factors that lead to high mobility in soils should correlate with a comparable ease of transfer from sediments into the boundary layer.

Transfer from the boundary layer into a biotic medium involves another set of considerations. Assuming that the bioconcentration factors (BCF) observed with fish reflect the net efficiency of this transfer process, several proven predictive tools, based on partitioning theory, become available, such as: log BCF (muscle) = 0.542 log P + 0.124 (13); log BCF (whole fish) = 0.85 log P − 0.70 (19). Since the kinetics of this process are directly related to the BCF, the modeling task should be relatively straightforward. But therein lies the rub. Those factors which lead to high mobility in soils usually lead to low bioconcentration potential in aquatic animals and vice versa (4). Consequently, quantification of bioavailability for purposes of refining environmental fate determinations will require an assessment of the "dynamic balance" existing between the abiotic and biotic "pools" contained in a system into which a chemical is introduced.

DISCUSSION

Sediment-water-chemical interactions that promote or impair bioavailability can be conveniently divided into three classes: physical, electrochemical, and biological. Obviously, these classes are interdependent, and each class can be further subdivided into different kinds of properties, conditions, or factors (Table 1). This type of classification is limited only by the imagination of the classifier. Each list of interactions has been restricted to three examples, because the intention is to illustrate how the quantification of bioavailability might be approached and not to provide a set of methods to refine environmental fate models.

Electrochemical Properties

Adsorption is generally regarded as a physical, surface-effect property. I have included it under the electrochemical group of interactions to emphasize that the bioavailability of those chemicals which are strongly adsorbed into soil or sediments as a result of electrostatic attractions, ion exchange, or chemisorption should be low. Chemicals that participate in these kinds of adsorption can often be identified as having a heat of adsorption greater than 10 kcal/mol (17). All other chemicals are probably adsorbed by more physical kinds of bonds that closely parallel a partitioning process (4) and will be considered later.

Because free energy concepts can be used to describe adsorption (4), it is tempting to draw analogies between bioavailability and enzyme kinetics. If the abiotic-biotic exchange system is assumed to be at equilibrium, then the available free energy content is zero. Conversely, a system at equilibrium can be brought to a state remote from equilibrium only if by some means, such as microbial activity, free energy can be made available (21). Then it becomes a question of how much free energy needs to be supplied to release an adsorbed chemical, which is analogous to the energy of activation. If the free energy required to reverse a sorption process exceeds some limit, reversibility by any biological means might be regarded as impossible. When this condition is observed it may be reasonable to assume that adsorption to sediments will serve as a permanent "sink."

The bioavailability of a chemical might depend on the formation of some type of intermediate complex between the transfer "agent" and the chemical. If so, bioavailability reaction rates might depend on the concentration of both the transfer agent and the chemical, in which case Michaelis-Menten kinetics, not first order, might apply. Experimentally, bioavailability reaction rates might be determined via the "back door." Consider, for example, a case where a microbial transformation or a fish bioaccumulation rate has been established in a liquid medium. If this rate were compared to that observed with a soil suspension, a relative measure of bioavailability might be derived. Thus, adsorption interactions affecting bioavailability may be amenable to modeling techniques analogous to those used to quantify enzyme kinetics.

TABLE 1. *Factors to consider when assessing the bioavailability of chemicals in aquatic environments*

Electrochemical properties	Physical conditions	Biological factors
Adsorption	Partitioning	Pelagic
Redox	Mass ratios	Benthic
pH-pK	Productivity	Microorganisms

Redox reactions may be one of the forgotten stepchildren of environmental chemistry when hazardous materials are considered. This is ironic because redox reactions are of paramount importance for understanding nutrient cycles and reactions at the mud-water interface in lentic environments (8). Conversely, these reactions are relatively unimportant in most river systems because oxygen is usually present at the interface. Perhaps one of the reasons redox reactions have not received the attention they deserve is the difficulty inherent in quantifying the role microorganisms play in these reactions. That is, microbes may create the redox conditions that lead to transfers or transformations, or a transformation process may change the redox potential.

The interplay between the environmental pH and the pK_a of a chemical should be a major consideration if the chemical has functional acidic or basic constituents. For example, maximum adsorption to soil and adsorption by biota would be expected where the environmental pH is roughly equivalent to the pK_a. Hence, if the pH of the sediment is different from that of the water column, shifts in chemical distribution and bioavailability could result that might not be intuitively obvious. It should also be recognized that redox reactions may be influenced by the pH of the medium or sediment (17).

Physical Conditions

Sometimes it seems that the world is one giant separatory funnel. Partition coefficients (log P) are used to predict such things as bioconcentration (13, 19), adsorption (4), biological membrane penetration (17), and evaporative losses (14). But is partitioning equally important in all these processes? If the "power" of the partition factor was greater for bioconcentration than for adsorption, the peculiar environmental problems associated with high log P chemicals, like DDT (dichlorodiphenyltrichloroethane) and some polychlorinated biphenyls, would be readily explained. But this is not the case. The regression equations between water solubility (log S) (where log P = 5.00 − 0.670 log S [9]) and soil adsorption normalized to organic carbon content (log K_{oc}) or bioconcentration factors for fish (log BCF) have nearly identical slopes (E. E. Kenaga and C. A. I. Goring, American Society for Testing and Materials Aquatic Toxicity Symposium, New Orleans, La., October 1978): log K_{oc} = 3.64 − 0.55 log S; log BCF = 2.791 − 0.564 log S. Thus, chemicals should be equally available to the abiotic and biotic components when partitioning determines their distribution. This also means that, with respect to distribution and bioavailability, chemicals which have low water solubility are not inherently more hazardous in the environment than more soluble chemicals.

These are rather surprising conclusions, because if all things were equal one would expect to find about the same concentration of a chemical in abiotic organic matter as in the biota. But all things are not equal. The mass ratio of sediments to water and the composition of solids participating in an exchange process vary widely with each kind of aquatic ecosystem. Also, similar kinds or quantities of solids may produce different outcomes in different kinds of systems. As a general rule, the relative bioavailability of chemicals for higher aquatic animals should increase as the depth and volume of a lake increases

(20) because the mass ratio between the sediments and the water decreases. Conversely, bioavailability should decrease as the turbidity of the lake increases, unless the chemical input is a result of erosional transport. For example, the concentration of DDT found in bass from nine Indiana lakes was negatively correlated with turbidity, primarily due to planktonic algae, whereas the concentration of dieldrin in bass from five reservoirs was positively correlated with turbidity, which was primarily due to suspended soil (18). Presumably the planktonic algae acted as an adsorbent to remove aerially introduced DDT residues from the water column, whereas the suspended soil served as a source of cyclodienes contained in runoff from treated crop land. Thus, the mass ratio as well as the composition of the solids may alter bioavailability of specific chemicals in entirely different ways in lakes and reservoirs.

A different outcome might be observed in clear versus turbid rivers. Turbidity in rivers should initially act like a "shock absorber" for a point source and then as a slow release mechanism as the flow proceeds downstream. This interaction will probably cause evaporative losses and photodegradation processes to be retarded (14), whereas bioavailability might remain relatively constant (15). For nonpoint sources, bioavailability per se should be a function of the concentration created by the mass loading equilibrated with the water flow and other "uncontaminated" solids in the system over time. The peak concentration from a nonpoint source will probably be less than what might be expected on the basis of soil residues because the greatest transport of chemicals adsorbed to soil generally occurs under flood conditions. Thus, the true concentration of chemicals in river water, and hence their bioavailability, may fluctuate widely because of fluctuations in river flow (6).

The productivity of an aquatic system is, of course, due to biological factors. Productivity is classified under physical conditions (Table 1) to emphasize the physical importance of planktonic algae and macrophytes in determining the bioavailability of chemicals to fish. Simple equilibrium considerations dictate that the relative concentration of a chemical in the water must decrease as the biomass, hence organic matter, in the system increases. Concurrently, the biomass and biotransformation activity of microbes should generally increase as the plant and animal biomass in an ecosystem increases (10). Thus, xenobiotics should pose fewer problems in eutrophic systems than in oligotrophic systems. Conversely, greater quantities of aquatic management chemicals may be needed in eutrophic systems than in oligotrophic systems (16) unless the product bioavailability can be "targeted" by innovative chemistry or application methods (3). Consequently, an assessment of bioavailability should include a negative term for productivity, which could probably be based on a phosphorus loading model (11), and a separate term for the mass ratio between solids and water (14).

Biological Factors

It must seem inconsistent that so little attention has been devoted to the recipients of biologically available chemicals: the macroorganisms. That is because, except for macrophytes, fish and invertebrates are just that, recipients. Their mass and activity are generally inadequate to direct the bioavailability of a chemical, although there may be some cases where the density of benthic invertebrates is great enough to promote sediment mixing (2). Yet it must be recognized that conducting a bioassay with a macroorganism is the best way to assess bioavailability. Chemical characterization and residue analysis of sediments do not provide a suitable basis for predicting the available forms or concentrations of contaminants. The analyst can only "ask" an organism what is available under a given set of circumstances and then extrapolate the results obtained to a wider range of conditions.

Bioassays could also be used to evaluate how benthic organisms obtain chemicals, whether by ingestion of contaminated solids or by direct uptake from interstitial water. For example, an experiment with treated sediment and water versus treated water and no sediment might demonstrate enhanced uptake by an organism (e.g., midge larvae) placed within the treated sediment. However, this evidence alone would not support the conclusion that uptake of the chemical resulted from ingestion of the treated sediment particles. The concentration of chemical in the interstitial water must then be determined (centrifugation is convenient) so that the uptake rate constants from water for both systems can be established. If the uptake rate constant from water for the treated sediment group is greater than the treated water group, ingestion probably contributed to uptake. If not, the depuration rate constants in both untreated sediment and water, including treatment group crossovers, should be determined. This is because the enhanced total uptake observed in the treated sediment group might be due to a retarded depuration rate resulting from the interstitial space surrounding the organism (i.e., the boundary layer) being quickly saturated. In other words, regardless of the outcome of an experiment to assess bioavailability, rate constants, or some other appropriate algorithm, should be determined, because without them extrapolation to natural environments will be subjective and tenuous.

The bioavailability of chemicals to microorganisms and how microorganisms influence the bioavailability of chemicals for macroorganisms are ambiguous subjects. Although I believe microorganisms rarely enhance bioavailability by direct mass transfers through a "food chain" (7, 12), their metabolic activities may frequently determine the bioavailability of a chemical to macroorganisms. Biotransformations alter bioavailability by changing the properties of chemicals or by changing the environment surrounding the chemicals (e.g., redox reactions). Evaluating these kinds of microbial activity in the laboratory is complex. Extrapolating the laboratory results and assessing their significance in natural environments is even more challenging. Nonetheless, as evidenced by this workshop, scientists are responding to these challenges and the actions of microorganisms that affect bioavailability should be better understood in the future, such that the whole science of hazard assessment will continue to mature.

CONCLUSIONS

The bioavailability of chemicals in aquatic environments depends on many interrelated factors. Foremost are the properties of the individual chemical.

These cannot be generalized, only measured and understood. Second are the properties of the receiving bodies. These can probably be quantified through application of the knowledge gained from studies on eutrophication. Finally, there are the properties and responses of the organisms. These may be at least partially derived from traditional aquatic toxicity studies, but there will always remain unknown qualities to be pondered, discovered, and enjoyed.

LITERATURE CITED

1. **Baughman, G. L., and R. R. Lassiter.** 1978. Prediction of environmental pollution concentration, p. 35–54. *In* J. Cairns, K. L. Dickson, and A. W. Maki (ed.), Estimating the hazard of chemical substances to aquatic life. ASTM Special Technical Publ. 657. American Society for Testing and Materials, Philadelphia.
2. **Boddington, M. J., A. S. W. deFreitas, and D. R. Miller.** 1979. The effect of benthic invertebrates on the clearance of mercury from sediments. Ecotoxicol. Environ. Safety 3:236–244.
3. **Florida Development of Natural Resources.** 1978. Aquatic weed identification and control manual, p. 99. Bureau of Aquatic Plant Research and Control, Florida Department of Natural Resources, Tallahassee.
4. **Freed, V. H., C. T. Chiou, and R. Hague.** 1977. Transport and behavior of chemicals in the environment—a problem in environmental health. Environ. Health Perspect. 20:55–70.
5. **Hamaker, J. W.** 1975. The interpretation of soil leaching experiments, p. 115–133. *In* R. Hague and V. H. Freed (ed.), Environmental dynamics of pesticides. Plenum Press, New York.
6. **Hamelink, J. L.** 1979. A proposed method for deriving effluent limits from water quality criteria. p. 127–131 *In* K. L. Dickson, A. W. Maki, and J. Cairns (ed.), Analyzing the hazard evaluation process. American Fisheries Society, Bethesda, Md.
7. **Hamelink, J. L., and A. Spacie.** 1977. Fish and chemicals: the process of accumulation. Annu. Rev. Pharmacol. Toxicol. 17:167–177.
8. **Hayes, R. F.** 1964. The mud water interface. Oceanogr. Mar. Biol. Annu. Rev. 2:121–145.
9. **Lambert, S. M.** 1968. Omega (Ω), useful index of soil sorption equilibria. Agric. Food Chem. 16:340–343.
10. **Lee, G. F., and A. F. Hoodley.** 1967. Biological activity in relation to the chemical equilibrium composition of natural water. Adv. Chem. Ser. 67:319–338.
11. **Lee, G. F., W. Rast, and R. A. Jones.** 1978. Eutrophication of water bodies: insights for an age-old problem. Environ. Sci. Technol. 12:900–908.
12. **Macek, K. J., S. R. Petrocelli, and B. H. Sleight III.** 1979. Considerations in assessing the potential for, and significance of, biomagnification of chemical residues in aquatic food chains, p. 251–268. ASTM Special Technical Publ. 667. American Society for Testing and Materials, Philadelphia.
13. **Neely, W. G., D. R. Branson, and G. E. Blau.** 1974. Partition coefficient to measure bioconcentration potential of organic chemicals in fish. Environ. Sci. Technol. 8:1113–1115.
14. **Southworth, G. R.** 1979. Transport and transformations of anthracene in natural waters, p. 359–380. ASTM Special Technical Publ. 667. American Society for Testing and Materials, Philadelphia.
15. **Spacie, A., and J. L. Hamelink.** 1979. Dynamics of trifluralin accumulation in river fishes. Environ. Sci. Technol. 13:817–822.
16. **Terriere, L. C., U. Kiigemagi, A. R. Gerlack, and R. L. Borovicka.** 1966. The persistence of toxaphene in lake water and its uptake by aquatic plants and animals. Agric. Food Chem. 14:66–69.
17. **Tinsley, I. J.** 1979. Chemical concepts in pollutant behavior, p. 265. Wiley-Interscience, New York.
18. **Vanderford, M. J., and J. L. Hamelink.** 1977. Influence of environmental factors on pesticide levels in sport fish. Pestic. Monit. J. 11:138–145.
19. **Veith, G. D., D. L. De Foe, and B. V. Bergstedt.** 1979. Measuring and estimating the bioconcentration factor of chemicals in fish. J. Fish. Res. Board Can. 36:1040–1048.
20. **Veith, G. D., and G. F. Lee.** 1971. Water chemistry of toxaphene—role of lake sediments. Environ. Sci. Technol. 5:230–234.
21. **White, A., P. Handler, and E. L. Smith.** 1959. Principles of biochemistry, 3rd ed., p. 1106. McGraw-Hill Book Co., New York.

Synopsis of Discussion Session

R. H. BRINK, D. BRANSON, AND G. BLAU

SUMMARY

1. Key factors in modeling the environmental exposure concentrations of a chemical include the use of estimation techniques for physical properties combined with predictions of environmental release. A material balance model can be used to estimate exposure concentrations for industrial workers, the public, disposal sites, and the environment.
2. There is an urgent need to develop physical property estimation methods based on constitutive relations for environmentally relevant properties when measured values are unavailable or questionable.
3. Material balance modeling can be used to analyze the importance of various environmental fate processes and to design relevant laboratory and field experiments, through sensitivity analysis. Also, information gained from material balance models should be an important element in decision making during a hazard assessment.
4. There was general agreement that only the amount of a chemical in solution is available for biotransformation by aquatic animals. For microbial transformations, evidence was presented that the amount of chemical in solution was important but may not be the only factor.

This section deals with two areas of concern in measuring or predicting the biotransformation and fate of chemicals in the aquatic environment. One area involves a material balance approach toward accounting for the distribution of a chemical following release into the environment. The second area is concerned with the bioavailability of a chemical released to an aquatic environment.

MATERIALS BALANCE APPROACH

The materials balance approach attempts to follow the fate of a chemical released to the environment in such a fashion that one can account for the entire mass of starting product.

One approach to a materials balance study is to conduct chemical analysis of various field samples in a well-planned monitoring program. This approach can be very useful in providing information on the environmental fate of chemicals which have been or are being released into the environment, and monitoring studies may be a necessary step in the validation of physical and mathematical models used to predict environmental fate. A properly conducted monitoring program must deal with chemicals which can be measured at low concentrations in environmental samples. Such studies are very expensive, and it is often necessary to make assumptions about the fate of significant fractions of the starting material not accounted for in the monitoring process.

Another approach is to conduct laboratory studies of varying complexity, accounting for the fate of the chemical by chemical or radioisotope techniques.

The environmental design may be relatively simple or quite complex. These studies are best carried out with appropriately labeled starting material and accounting for the distribution of the labeled molecules at some appropriate time after introduction of the test material. It seems reasonable to predict that the more complex assemblages will be most useful in validating mathematical models or other predictive techniques.

A third approach is to attempt to predict the environmental fate of a chemical by using available data on the physical and chemical properties of the chemical and information on its production, distribution, use, and disposal, including all real or potential releases to the environment and significant characteristics of the environment to which the chemical may be released. This approach, as used by one industry and discussed in detail in the first paper of this session, is accepted as a valuable, and perhaps necessary, early step in the hazard evaluation of chemicals released into the environment. After the collection of information on the manufacture, distribution, use, and environmental release (including disposal) of the chemical, it may be possible to arrive at many conclusions with respect to the fate of the chemical in a specific environment, especially on the transport possibilities of that chemical, using readily obtainable physical and chemical data to estimate those physical parameters of environmental relevance. This approach may serve several functions, as demonstrated in the discussion paper, and this process may serve to identify the relative importance of factors, such as whether the reduction of certain contaminants in a product produces any significant advantages with respect to the reduction of hazard. For predicting the environmental fate of new or existing chemicals, the use of estimation techniques to provide values for parameters such as volatility, octanol-water partition coefficient, soil sorption, and the like can provide early and cost-effective information to identify the most important factors and data needs. A sensitivity analysis, wherein one follows the extent of variation in the output conclusions as a result of variations in each input parameter, varied one at a time, is used to identify the important factors. To the extent that each estimation step is well defined as to the precision or certainty of the prediction, for specific classes of chemicals, it may be possible to forego laboratory evaluations of certain parameters. The statistics involved in the prediction techniques are very important and the uncertainty of the data must be reflected in any conclusions. Also, insofar as possible, the estimations should be based on values obtained under environmentally relevant conditions; for example, the volatility of a chemical from a water body may be estimated, given a knowledge of the vapor pressure, water solubility, and sorptive tendencies of the chemical. But the values for vapor pressure, water solubility, and sorption should be valid for ambient temperature, pH, and so forth, because extrapolations from nonambient conditions can provide erroneous information. It is also important to realize that most of the techniques for estimating physical and chemical parameters are good only for nonelectrolytes and that the principal application is in estimating how the starting material may partition throughout the environment without providing information on rates. Despite these limitations, this approach can provide valuable insight on the fate of a chemical in a given environment and on environmental exposure concentrations. One of the most important potential uses is in the identification of the most critical parameters in the environmental fate of the particular compound to indicate what laboratory testing or monitoring studies might be necessary and which might reasonably be avoided.

BIOAVAILABILITY

The bioavailability of a chemical may be concerned with the extent to which a living organism or community of organisms may be able to sorb, store, or transform that chemical through biologically active, enzyme-mediated processes or the extent to which the chemical may alter the normal functions of an organism or a community or organisms. Bioavailability implies a spatial relationship which brings the organism(s) and the chemical sufficiently near to each other for the initiation of some activity and undoubtedly is related to such factors as the form of the chemical (e.g., dissociated versus undissociated) and the physiological conditions of the organism. Operationally, for the purposes of this workshop, we will define bioavailability as the extent to which a chemical compound in an aquatic environment is available for biologically mediated transformations such as degradation by microorganisms, uptake by plants, and storage of metabolites by aquatic animals.

In this discussion paper we were presented with considerations of the factors which determine the presence of a chemical in the water phase adjacent to the organism of concern. Water solubility is a principal factor in allowing a compound to become available to organisms above the microbial level in the sense of a physical presentation of the chemical to the membrane surfaces. That this is not the only important factor in bioavailability is evident in the phenomenon of bioaccumulation of chemicals by aquatic organisms, with the bioaccumulations of greatest concern being those of relatively water-insoluble compounds in the fatty tissues of the organism. But for bioavailability of a compound for metabolism or other transformations by aquatic plants or animals, the tendency of the compound to exist largely as a solute in water is probably the most important factor with respect to that availability. Likewise, a chemical which is strongly sorbed to the sediment or to suspended matter will be relatively unavailable to such organisms.

The situation with respect to microbial transformations is not as clear. It cannot be assumed, a priori, that the sorption will or will not make a chemical less available for transformation by microorganisms. In some cases, the chemical may be trapped between clay layers or in interstices of the particulate matter and, as a result, may be physically unavailable to microbial attack. In some instances the intimate assemblage of microorganisms and organic substrate, both sorbed and inert and suspended particulate matter, can lead to biodegradation rates substantially greater than the rate one would have observed if the microbial cell were freely distributed throughout the system. It may be that in the microenvironment at the particulate surface, there is a need for a water boundary layer between the substrate and the microbial cell wall, but there was disagreement on whether or not the substrate must be in true solution in the boundary water before sorption and metabolism can occur.

It is recognized that there are many other factors which can be important in the bioavailability of chemicals in the aquatic environment. Among these factors are:

1. The transformation of the chemical by abiotic processes which may render the chemical more or less available for biotransformation
2. The diffusion of the chemical through microbial aggregates to actively metabolizing cells or through sediments. This mass transfer process can be a rate-limiting step.
3. The need for extracellular enzyme processes to initiate attack on bulky, undissolved molecules such as cellulose
4. The presence of other metabolizable organic matter which may be necessary to provide for the active growth of microorganisms when there is a need for enzymatic adaptations to occur before the organisms can transform the test chemical
5. The presence of the compound at such low concentrations that it persists either because its concentration is below some threshold required for enzymatic processes or because the reaction rate becomes extremely slow at very low substrate concentration, giving the appearance of unavailability

The extent to which bioavailability affects or alters a determination of existing environmental concentration for a test compound is a rather complex question that needs further study. Bioavailability to higher organisms, for example, is more important in the consideration of effects on those organisms than on environmental concentration. Undoubtedly uptake by plants and animals has some effect on the environmental concentrations of a compound, but for the vast majority of organic compounds transported in the aquatic environment, microbial transformation is the overwhelming factor in the reduction of their concentration in the environment. That this is so is evident from the large biomass of microorganisms and their very rapid metabolic rates. There is a need, however, to learn more about the significance of uptake by plants and animals in the removal of specific compounds from aquatic environments.

Chapter 5. ENVIRONMENTAL EXTRAPOLATION OF BIOTRANSFORMATION DATA

Role of Biodegradation Kinetics in Predicting Environmental Fate

R. J. LARSON

Environmental Safety Department, Procter and Gamble Company, Cincinnati, Ohio 45217

Predictions of environmental concentration are key to assessing the environmental hazard associated with the use of specific organic chemicals. Inherent in the prediction of environmental concentration is the identification and characterization of formation and degradation processes which control the steady-state levels of a chemical. In general, formation and degradation processes are difficult to study in the field, and model systems must usually be set up in the laboratory to study a particular compound. The difficulty then comes in extrapolating laboratory results to the field situation and in gauging the accuracy of these extrapolations. Steady-state levels can be predicted from laboratory studies by determining the rate of important chemical and biological reactions. Whereas first-order rate expressions have been found to describe a variety of chemical reactions, a more complicated situation exists for biodegradation reactions. The kinetics of ultimate biodegradation were studied in dilute assay systems typical of soils and surface waters. The rate equation developed to describe biodegradation was found to contain first-order, mixed-order, and zero-order regions, depending on the compound used and the character of the microbial population. In many cases, however, the rate of biodegradation could be approximated by a first-order equation, and extrapolations could be made by using first-order rate constants. The effect of population size and diversity, temperature, and other metabolic variables could be incorporated into biodegradation rate expressions to cover a variety of environmental situations.

It is generally accepted that estimates of environmental concentration are a key element, along with aquatic effect levels, in predicting the environmental hazard associated with the use of specific organic chemicals (1). Accurate predictions of environmental concentration require that the chemical and biological fate of a particular material be estimated for a variety of environmental compartments. Environmental systems are composed of many environmental compartments, each of which interacts with the others in a complex and dynamic equilibrium. Since it is very difficult to control or even identify all of the important variables within multiple environmental compartments, model systems are usually set up in the laboratory to simulate a particular compartment. As a result, it is often necessary to extrapolate from laboratory data to predict environmental hazard and to estimate how accurately these extrapolations reflect the situation in the "real world."

The extrapolation of laboratory results to actual environmental systems requires the identification of two major pieces of information: (i) the most important compartments involved in the transformation of a specific chemical and (ii) the most important biological or chemical mechanism in a particular compartment responsible for the transformation. In this context, "most important" refers to that compartment or mechanism responsible for the greatest net increase or decrease in the level of a material. To accurately estimate net changes in the level of chemicals, it is necessary to define the rates at which various formation and decomposition processes occur and to rank these rates into some sort of hierarchy. Those reactions which occur at the fastest rate will be most important in controlling the fate of a chemical in a particular compartment. Those which occur at the slowest rate will be the least important, *unless* they can be shown to be rate limiting for other chemical or biological reactions. To develop this hierarchy of rates approach, suitable kinetic expressions must be found to define the rate of appropriate chemical and biological reactions. To ensure the practicality of this approach, only those variables which affect reaction rates most can be considered.

Kinetic expressions have been developed to describe the rate of various chemical reactions in environmental systems (1), specifically those involving hydrolysis, photolysis, ionization, vaporization, and physical/chemical sorption processes. Studies of the kinetics of biological reactions have been less extensive, being mainly confined to uptake kinetics of natural compounds in various environmental systems (12–15) or to activity measurements of indigenous microflora. Effort has not been directed toward developing rate equations to describe biodegradation or biometabolism of xenobiotic organic pollutants in environmental systems, or toward identifying the important variables which significantly affect the rates of biotransformation in these systems. However, it is exactly this information which is necessary for extrapolating laboratory rate data to actual environmental systems. To develop this area further, this paper will concentrate on two major areas: (i) the types of kinetic equations appropriate for predicting rates of biodegradation in dilute environmental systems (i.e., those systems like soil and surface waters where the levels of chemicals and biomass are low, as opposed to more "concentrated" systems like wastewater treatment where these levels are high), and (ii) identifying and characterizing those variables which significantly affect rates of biodegradation and are therefore most important in extrapolating laboratory rate data to the environment.

KINETICS OF BIODEGRADATION

In general, rates of biological reactions tend to be hyperbolic saturation functions of substrate concentration. The kinetics of bacterial growth and many enzyme reactions have been found to follow the general equation

$$r = R_{max}S/(K_s + S) \tag{1}$$

where r is the reaction rate, R_{max} is the maximum reaction rate, K_s is the half-saturation constant where $r = 0.5R_{max}$, and S is the substrate concentration. A graphical representation of equation 1 is shown in Fig. 1. If equation 1 is integrated:

$$R_t = K_s \ln \frac{S_0}{S_t} + (S_0 - S_t) \tag{2}$$

where S_0 is the initial substrate concentration. It is evident from equation 2

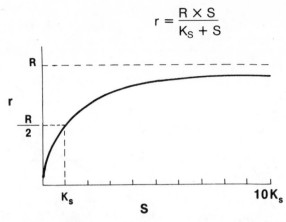

$$r = \frac{R \times S}{K_S + S}$$

FIG. 1. *Plot of reaction rate, r, as a function of substrate concentration, S, for reactions following the equation:* $r = R_{max}S/(K_s + S)$.

and Fig. 1 that the rate equation describing r as a function of S is a fairly complex one and contains first-order, mixed-order, and zero-order regions. When $S \gg K_s$, $K_s + S \cong S$, and the reaction approaches zero order with

$$r = R_{max} \tag{3}$$

and R_{max} becomes the limiting maximum reaction rate. At $S = 10K_s$, the deviation from zero order is $<10\%$. When $S \ll K_s$, equation 1 reduces to

$$r = R/K_s \times S$$

and the reaction approaches first order with R/K_s = first-order rate constant. Strictly speaking, the reaction rate approximates first order only when $S < 0.1K$ and deviations are $<10\%$. Due the hyperbolic nature of the saturation curve, however (Fig. 1), r does not vary significantly with S until K_s is exceeded. Practically speaking, therefore, approximate first-order kinetics can be assumed for all values of $S < K_s$ (Fig. 1). At all other values of S up to the zero-order region, r is mixed order and determined solely by the two constants R and K_s.

Equation 1 has typically been used to follow uptake kinetics of natural products in environmental systems (12–16). It can also be used to follow degradation kinetics in these systems. As indicated in Fig. 2, the rate of CO_2 evolution during glucose degradation in a typical Ohio River water sample is also a saturable function of glucose concentration with $K_s = 26$ mg/liter. Practically speaking, first-order kinetics are observed at glucose concentrations approaching the K_s value since the CO_2 data can be fit with good correlation ($R^2 = 0.97$) to a first-order equation when $S = 20$ mg/liter (Fig. 2).

Since reaction rates (r) are a hyperbolic saturation function of S, the constants R and K_s are very important in determining the order of biological reactions following equation 1. They are specific for a particular bioassay system and allow the order of the reaction rate to be determined at any substrate con-

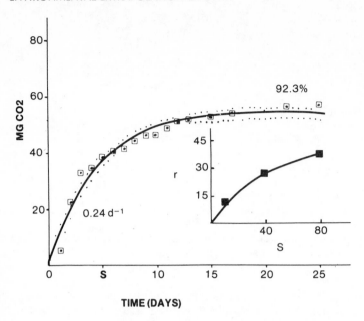

FIG. 2. *Kinetics of glucose mineralization in Ohio River water. D-Glucose was tested at various concentrations, and the initial rate, r (mg of CO_2 per day), was plotted as a function of substrate concentration, S (mg/liter; see insert). Lineweaver-Burk replots were made and analyzed via linear regression with K_s = 26 mg/liter, R_{max} = 17 mg of CO_2/liter × days (R^2 = 0.999). Glucose degradation followed approximate first-order kinetics at glucose concentrations approaching K_s (□, 20 mg/liter), and could be fit to a first-order equation $y = a(1 - e^{-bt})$ where a = extent of degradation (%) and b = rate constant (day^{-1}). Dotted contours are 95% confidence limits of the true mean.*

centration. Substrate concentration, therefore, becomes a key variable in conducting biodegradation assays and in extrapolating biodegradation rate data from the laboratory to the real world. In practice, observed rates of biodegradation are almost always proportional to substrate concentration. The question of prime importance is what type of rate constant, k, should be used to describe this proportionality in the environment and what is the most environmentally relevant rate function, r, to be measured.

Kinetic models to predict rates of biodegradation have typically defined r as a growth function and have relied on the Monod equation (6) to describe microbial growth. Growth measurements are actually indirect measures of microbial degradation in that they follow the rate of increase in microbial numbers as a function of substrate concentration. Environmentally, we are more concerned with determining the rate of decrease of a compound or the rate of increase in the level of a metabolite. While growth and degradation are obviously related, correlation of the two requires that certain assumptions be made to translate increases in biomass to degradation of material. One of these assumptions, that the growth yield Y is constant, equal to 0.5 (50%), is not applicable to dilute environmental systems. This is indicated by results of CO_2 evolution studies conducted in Ohio River water samples using a [^{14}C]ethoxylate-labeled $C_{12}E_9$ alkyl ethoxylate (Fig. 3). Incorporation of disintegrations per minute into the >0.45-μm particulate fraction did not exceed 10%, and 90% CO_2 conversion efficiencies were observed from a concentration range of 50 to 10,000 μg/liter. The 10% incorporation figure must be considered a maximum level since it also includes disintegrations per minute sorbed to the sediment fraction present in the water sample. These results indicate that significantly more than 50% of the carbon in a compound can be catabolized to CO_2 in dilute environmental systems (as a result of maintenance energy requirements) (5) and that growth, as defined by the Monod equation, is not a requirement for degradation.

Accurate measurements of microbial growth are difficult in environmental systems, especially when the concentration of chemicals is low. For example, even at 10 mg/liter (which would be a high concentration in the environment), a good carbon source can only generate about 10^6 cells per ml, assuming an óptimistic growth yield, $Y = 0.5$, and average dry cell weights of 10^{-12} to 10^{-13} g per cell. It is difficult to accurately measure the rate of change of viable cells as a function of S at these levels, or to ensure that all microbial types in an environmental sample are growing on the laboratory media used. Fluorescence techniques can be used for direct counts, but enumeration of bacterial numbers does not take into account bacterial activity, which recent evidence (13) indicates has a substantial effect on metabolism of natural organics in environmental systems. A truly integrated approach to measuring microbial growth would have to incorporate both growth and activity measurements, and would have to be applicable to all types of environmental compartments—

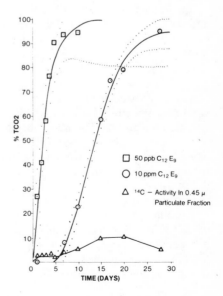

FIG. 3. *Distribution of radioactivity during mineralization of ^{14}C-labeled $C_{12}E_9$ alkyl ethoxylate (AE) in Ohio River water. The percentage of radioactivity evolved as CO_2 is shown for 50 ppb (□) and 10,000 ppb (○). Incorporation of disintegrations per minute into the >0.45-μm particulate fraction is shown for the 10,000-ppb concentration (△). The CO_2 data for 50 and 10,000 ppb have been fit by computer to equations 6 and 8, respectively, using iterative techniques. Dotted contours are 95% confidence limits of the true mean.*

□ 50 ppb $C_{12}E_9$

○ 10 ppm $C_{12}E_9$

△ ^{14}C − Activity In 0.45 μ Particulate Fraction

% TCO2

TIME (DAYS)

surface waters, soils, and sediments. Autoradiography studies (8) using radiolabeled chemicals are one possibility, but these studies would be time-consuming and expensive, and from a practical standpoint unnecessary, since environmental degradation studies are concerned with measuring how fast microorganisms effect changes in the level of chemicals, not vice versa.

If we accept the fact that microbial growth kinetics cannot be used to predict biodegradation rate constants on a routine basis (although some excellent work has been done in this area [6]), what other rate functions can be measured? The simplest and most direct approach is to measure the rate of decrease in the level of a chemical as a function of substrate concentration, holding other variables constant. The decay curve generated in this fashion can often be fit to a first-order·equation of the form

$$y = ae^{-bt} \qquad (4)$$

where y is the level of material remaining, a is the level at time zero, b is the rate constant (time^{-1}), and t is time. The first-order rate constant, b, can be determined directly via a computer, using nonlinear regression techniques, or by linear regression, using least-squares analysis. In the latter case, equation 4 must be converted to

$$\ln y = \ln a - bt \qquad (5)$$

where the rate constant, b, is the slope of the resulting straight line. A modified form of equation 4 can also be used to follow product formation during degradation of a specific chemical:

$$y = \begin{cases} 0 & \text{for } t \le c \\ a\{1 - \exp[-b_1(t-c)]\} & \text{for } t > c \end{cases} \qquad (6)$$

where y is product formed, t is time, a is the extent of degradation (asymptote), b_1 is the rate constant (time^{-1}), and c is the lag time prior to degradation.

Equation 6 basically describes the three regions of a microbial product curve and allows both the extent of degradation and the lag time prior to degradation to be estimated. In a modified form, it can also be used to follow a decay function:

$$y = (100 - a) \exp[-b_1(t-c)] + a \qquad (7)$$

Carbon dioxide evolution during degradation of a variety of chemicals in laboratory systems can be accurately described by a first-order expression like equation 6 (Table 1). The rate constants generated for a particular compound do not show a pronounced difference when different inoculum sources are used (Table 2, Fig. 4) or when population densities vary from 10^3 to 10^6 colony-forming units (CFU)/ml, a range typical of surface waters (6, 13). Equation 6, therefore, provides a statistical basis for analyzing biodegradability data and comparing compounds between experiments.

It should be mentioned that the observed first-order rate constant b_1 is an overall value for a number of metabolic steps. Due to analytical constraints, it is generated using relatively high concentrations ($\sim$10 mg/liter) of test material and may underestimate the true b. There is also good evidence that discrete alterations of some chemicals (pesticides) in surface water samples

TABLE 1. *Rate constants for mineralization of various compounds in synthetic media*

Compound	b (days^{-1})	$R^{2\,a}$
Linear alkylbenzene sulfonate	0.10	0.995
Nitrilotriacetate (sodium salt)	0.30	0.998
$C_{14}E_7$ alkyl ethoxylate	0.13	0.999
$C_{13}E_{6.5}$ alkyl ethoxylate	0.13	0.950
Cetyldimethylpolyoxyethylene sulfonate	0.07	0.993
Alkyl dimethylamine oxide	0.07	0.990
Hexadecyltrimethyl ammonium chloride	0.20	0.992
Hexadecylsulfonate	0.20	0.992
Isopropylpalmitate (insoluble)	0.27	0.980
Glucose	0.20	0.995

[a] Correlation coefficient for fit of CO_2 evolution data to equation 6.

at environmental concentrations do show a more pronounced dependency on population size so that b is really a pseudo-first-order or second-order rate constant (1). As we shall see later, pseudo-first-order kinetics are also observed during degradation of some chemicals in the laboratory. For the most part, however, ultimate degradation of many chemicals in both terrestrial and aquatic test systems can be described by first-order equations similar to equation 6. Since mineralization studies can often be conducted without tagged material or specific analytical methods, it is convenient to use kinetics of this type as a first approximation so that rate constants for a variety of chemicals can be estimated.

Pseudo-first-order kinetics are sometimes observed during degradation of chemicals in the laboratory. This is indicated by the S-shaped character of the degradation curve (Fig. 5). The equation used to describe a product curve of this type is

$$y = a\{1 - \exp[-b_2(t - c)^2]\} \qquad (8)$$

TABLE 2. *Biodegradation of linear alkylbenzene sulfonate and glucose by different inocula*

Inoculum	Compound	Initial population density (CFU/ml)	a (% total CO_2)	b (days^{-1})	c (days)	R^2
Activated sludge	LAS	1.2×10^4	77.5	0.10	5.5	0.995
	Glucose	1.2×10^4	89.0	0.20	0	0.980
Influent wastewater	LAS	6×10^6	79.1	0.13	2.7	0.994
	Glucose	6×10^6	89.0	0.40	0.2	0.983
Influent wastewater	LAS	5.4×10^3	78.9	0.12	3.6	0.998
	Glucose	5.4×10^3	91.4	0.23	0	0.985
Effluent wastewater	LAS	1.9×10^4	73.2	0.12	4.6	1.000
	Glucose	1.9×10^4	90.0	0.20	0	0.996

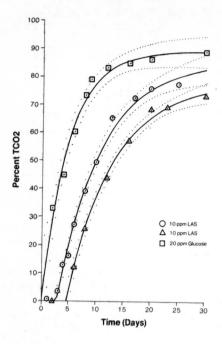

FIG. 4. *Effect of population size on kinetics of CO_2 evolution during degradation of LAS (10 mg/liter). The CO_2 evolution data have been analyzed by equation 6. Symbols:* △, *activated sludge inoculum, 1.2×10^4 CFU/ml;* ○, *influent sewage inoculum, 6.0×10^6 CFU/ml;* □, *glucose control (20 mg/liter), with activated sludge inoculum. Dotted contours are 95% confidence limits of the true mean.*

where b_2 is the rate constant (day^{-2}). The decay curve follows the form

$$y = (100 - a) \exp[-b_2(t - c)^2] + a \qquad (9)$$

The main difference between equation 8 or 9 and the standard product or decay curve (equation 6 or 7) is in the exponential character of the initial uptake phase. The biological significance of this uptake phase is most easily understood by going back to the original differential equations, which are, respectively

$$dy/dt = k \cdot y \qquad (10)$$

and

$$dy/dt = k \cdot ty \qquad (11)$$

Equation 10 indicates that the rate of change of y is only a function of the concentration of y, whereas in equation 11 this change is initially a function of both y and t. Time, t, can be regarded most simply in biological terms as the time necessary to build up a competent bacterial population (acclimation period). In essence, then, it represents a concentration function, i.e., the concentration of bacteria, and dy/dt is actually a second-order reaction in the initial part of the degradation curve since it is a function of two concentration terms, y and t. Strictly speaking, therefore, b_2 is actually a second-order constant until the point of inflection of the S-shaped curve is reached. When this inflection point is reached (and it has to be reached in batch systems), the concentration of bacteria is no longer rate limiting, and substrate once again controls the reaction rate. The reaction then becomes a first-order function of y, and b_2 is a first-order rate constant.

The above kinetic discussions are best illustrated in the laboratory with re-spike experiments. If an S-shaped degradation curve is observed upon initial exposure to a material, reflecting a buildup in the level of competent bacteria, then subsequent respikes to that established population should result in stand-ard decay or product curves described by equation 6. This response is shown for an amine oxide in Fig. 5. Upon initial exposure of microorganisms to alkyl dimethylamine oxide, an S-shaped degradation curve was observed. Re-exposure of these microorganisms to alkyl dimethylamine oxide resulted in a standard first-order product curve with rate constant b_2 comparable to b_1 after normalization to the same units (days^{-1}) (Table 3). In both cases, high CO_2 conversion efficiencies (90%) were observed, which indicates that the increase in cell numbers during the exponential uptake phase does not sub-stantially increase cell yield coefficients; i.e., Y is still <0.1.

Pseudo-first-order degradation kinetics have been observed for minerali-zation of other chemicals. Tiedje and Mason (11) have shown that degradation of nitrilotriacetate (NTA) in soil follows an S-shaped pattern upon initial ex-posure of soil microorganisms to NTA. Subsequent respike curves follow the standard pattern, similar to the results reported here for alkyl dimethylamine oxide. An S-shaped curve is also observed when mineralization studies are conducted in environmental water samples with the use of relatively high con-centrations of test material (Fig. 3). Results of this type indicate that metabolism of a chemical can be mediated by microorganisms already present in the en-vironmental sample, although not at a rate dictated by the concentration of the chemical.

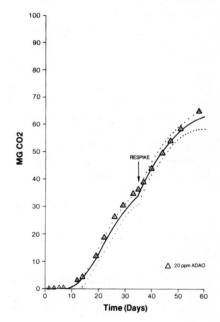

FIG. 5. *Kinetics of CO_2 evolution dur-ing degradation of alkyl dimethylamine oxide (ADAO) at 20 mg/liter. After 35 days (indicated by arrow), an additional 20 mg/liter respike was made. The CO_2 data were analyzed as described in Table 3. Dotted contours are 95% con-fidence limits of the true mean.*

TABLE 3. *Biodegradation of alkyl dimethylamine oxide with multiple exposures*

Expt	a (% total CO_2)	b_2 (days^{-2})	$\sqrt{b_2}$ (days^{-1})	b_1 (days^{-1})	c (days)	R^2
Initial[a]	96.3	0.0032	0.056		7.6	0.99
Respike[b]	91.8			0.063	1.0	0.99

[a] The CO_2 data were fit to equation 8.
[b] The CO_2 data were fit to equation 6.

Although there are certain advantages to assuming first-order degradation kinetics in laboratory systems, a more rigorous theoretical basis for extrapolating laboratory rate constants to the environment is highly desirable for a number of reasons. First, rate constants for degradation of certain chemicals, i.e., pesticides, clearly are dependent on population size, and a second-order rate expression is necessary to normalize observed rate constants for bacterial density or "concentration" (D. F. Paris and W. C. Steen, U.S. Environmental Protection Agency, personal communication). Normalization for bacterial numbers, however, does not necessarily take into account bacterial activity, which can be important in determining overall rates of degradation (13). In addition, we have seen that ultimate degradation of many chemicals is first order with respect to substrate concentration and that population density, within the range found in surface waters, does not play as important a role. Clearly, a theoretically relevant "middle ground" is necessary to assess the effect of both substrate levels and population size on biodegradation rate constants and to facilitate the extrapolation of these rate constants from the laboratory to the environment. That middle ground may well involve measurements of bacterial activity.

It has been shown that glucose mineralization in environmental freshwater samples can be described by equation 1. Wright (12) has shown that similar kinetics are observed for glucose mineralization in marine systems. Not surprisingly, therefore, equation 1 can also be applied to degradation of xenobiotics in laboratory systems. For example, the rates of mineralization of two quaternary surfactants, alkyl dimethylamine oxide and alkyl trimethyl ammonium chloride, are both saturable functions of substrate concentrations (Fig. 6 and 7). Initial velocities can be determined for these chemicals at various concentrations, and Lineweaver-Burk replots can be made via linear regression to yield R_{max} and K_s values. The R_{max} and K_s can also be determined graphically from one concentration by using equation 2 and plotting

$$\frac{2.3}{t} \log \frac{S_0}{S_t} = \frac{-1}{K_s} \frac{(S_0 - S_t)}{t} + \frac{R_{max}}{K_m} \tag{12}$$

Equation 12 is in the form of a straight line with slope $= -1/K_s$ and y intercept R_{max}/K_s.

Irrespective of how K_s and R_{max} are determined, these constants uniquely define the rate equation for mineralization of a specific compound. The ratio R_{max}/K_s also represents the first-order rate constant for degradation when $S < K_s$. This rate constant incorporates both the activity of the degrading pop-

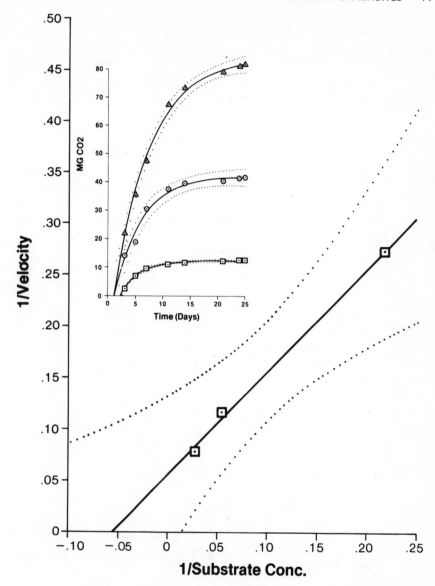

FIG. 6. *Kinetics of CO_2 evolution during degradation of alkyl trimethyl ammonium chloride at 5 (□), 20 (○), and 40 (△) mg/liter (inset). The initial rates of CO_2 evolution were determined by taking the first derivative of equation 6 at t = c. Lineweaver-Burk plots were made and analyzed by linear regression, R^2 = 0.997. Dotted contours are 95% confidence limits of the true mean.*

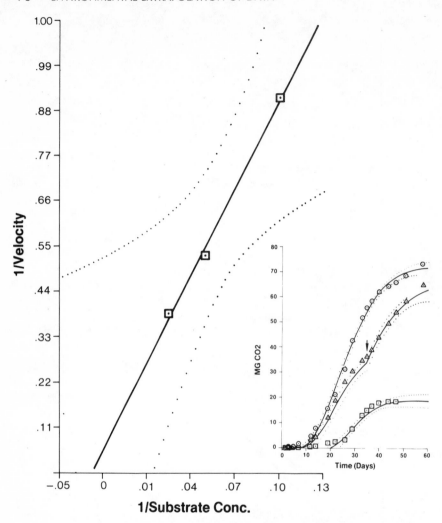

FIG. 7. *Kinetics of CO_2 evolution during degradation of alkyl dimethylamine oxide at 10 (□), 20 (△), and 40 (○) mg/liter (inset). The initial rates of CO_2 evolution were determined by taking the first derivative of equation 8 at the point of inflection of the S-shaped curve. Lineweaver-Burk plots were made and analyzed by linear regression, $R^2 = 0.996$. Dotted contours are 95% confidence limits of the true mean.*

ulation and the substrate dependency of the reaction. It therefore takes into account both population and substrate levels, and provides a theoretical basis for extrapolating laboratory rate data to the environment.

Calculated R_{max} and K_s values for degradation of alkyl trimethyl ammonium chloride, alkyl dimethylamine oxide, and various chemicals in different laboratory and environmental systems are shown in Table 4. It is evident that, although observed first-order rate constants are a reasonable approximation of the R_{max}/K_s values, they tend to *underestimate* the true values since they are not obtained at a substrate concentration where the reaction rate is, strictly speaking, first order. Nevertheless, laboratory values err on the conservative side, and calculated R_m/K_s values can be used to predict first-order rate constants when $S \ll K_s$, as will be the case in most environmental systems. First-order rate constants are particularly useful since they allow half-lives for chemicals to be estimated by the following:

$$0.693/k = t_{1/2} \tag{13}$$

The turnover time or rate can also be determined if the background concentration of the compound is measured for a particular environmental system. Both half-life and turnover time are important environmental parameters since they allow flux rates to be calculated for a material.

As shown in Table 4, R_{max} values are related to microbial population densities when these densities are high. The R_{max} for NTA degradation in activated sludge is at least 10-fold higher than the R_{max} in river water, even though K_s values remain fairly constant. This high value reflects the very high biomass levels in wastewater treatment systems, where observed rate constants are usually linear functions of biomass levels. On the other hand, R_{max} values for NTA degradation in different dilute systems show much less variation, reflecting the low biomass levels in these systems. It can be postulated that R_{max} values for degradation are not significantly affected until a certain biomass concentration is reached, possibly $>10^7$ to 10^8 CFU/ml.

EFFECT OF VARIABLES ON BIODEGRADATION RATE CONSTANTS

In the previous section, the types of kinetic equations appropriate for generating rate constants for biodegradation were evaluated. In this section, attempts will be made to characterize those biological variables which significantly affect these rate constants and their extrapolation to environmental

TABLE 4. *Kinetic constants for biodegradation in various systems*

Compound[a]	System	Observed k (day^{-1})	R_{max} (ppm/day)	K_s (ppm)	R_{max}/K_s (days^{-1})	R^2
NTA	Ohio River water	—	46	110	0.42[b]	0.999
NTA	Synthetic medium	0.30	39	114	0.34	0.994
NTA	Semicontinuous activated sludge	—	360	81	4.4[b]	0.999
NTA	Soil	—	81	368	0.22[c]	0.998
ADAO	Synthetic medium	0.06	6	33	0.18	0.996
ATAC	Synthetic medium	0.30	18	18	1.0	0.986
Quinoline	Batch fermentation	—	179	29	6.2[d]	0.991
Glucose	Ohio River water	0.24	17	26	0.65	0.999
NTA	Soil	0.04	—	—	—	0.970

[a] NTA, Nitrilotriacetate; ADAO, alkyl dimethylamine oxide; ATAC, alkyl trimethyl ammonium chloride.
[b] Calculated from Thompson and Duthie (10).
[c] Calculated from Tiedje and Mason (11).
[d] Calculated from Smith et al. (6).

systems. This discussion will primarily cover those variables which directly affect microbial metabolism and/or can be expected to be rate limiting under various conditions. Variables which indirectly affect metabolism by controlling the "biologically available" concentration of a material (for example, sorption/desorption processes) will not be discussed, although they may be rate limiting under certain conditions. Many of these chemical/physical variables have already been considered elsewhere (1) or will be examined later in this volume.

Inoculum

The composition of the microbial inoculum probably represents the single most important variable in biodegradation studies. Regrettably, in comparison to other variables like substrate concentration, it is also the one over which environmental control is most limited. Degradation studies have traditionally been conducted with pure cultures isolated via selective enrichment techniques. More recently, artificially constructed mixed cultures have been used to examine competitive or synergistic effects. These techniques lead to a reduction in microbial diversity and simplify experimental procedures in the laboratory. Unfortunately, when the results of laboratory studies are extrapolated to the environment, questions about the environmental significance of the data invariably occur.

The use of metabolically diverse mixed microbial cultures decreases the difficulty in extrapolating laboratory data to the environment. Since the microbial diversity of even dilute systems ($\sim 10^6$ CFU/ml) is subject to considerable variation in different water samples, development of a "standard" inoculum is not practical for most studies. The only parameter that can reasonably be controlled is total cell numbers. Ideally, the best aerobic system from which to select a population of a given size would be a continuous one operating at more or less steady-state conditions. The system would be exposed to a variety of different organic compounds at realistic concentrations (micrograms per milliliter range) and would contain a relatively high population of diverse microbial types. Acclimation conditions in this system would be rigorous enough to allow selection of competent bacteria, but not so severe that selective enrichment would be a problem. Fortunately, such a system does exist and has considerable environmental relevance since it is used as a form of secondary wastewater treatment. The system is activated sludge and its laboratory counterpart is the semicontinuous activated sludge (SCAS) system.

Activated sludge is composed of a variety of different microbial types and represents a scientifically and environmentally relevant source of mixed microbial cultures (3, 4). Batch activated sludge systems (SCAS systems) can easily be set up in the laboratory (7) to generate high levels of diverse microorganisms (2,000 to 3,000 mg/liter of mixed liquor suspended solids). In SCAS systems, microorganisms are exposed to test materials in the presence of alternate nutrients so that growth and/or adaptation of competent bacteria, if necessary, can occur. This feature is important since acclimation conditions are not present in dilute batch systems used for biodegradation assays. Net growth is minimal (Fig. 3), population levels and diversity are low, and steady-state conditions cannot be reached without a number of serial transfers as a result of low microbial growth rates. Actually, biodegradation assays in dilute systems primarily measure the "performance" of the existing population. Attempts to develop an acclimated population, therefore, must be initiated prior to the start of a degradation study, not during it.

Activated sludge cultures can be used to generate reproducible numbers of diverse microorganisms for biodegradation assays. As shown in Table 5, the number of CFU per milliliter in the supernatant of homogenized activated sludge generated in an SCAS system is fairly constant between experiments, ranging from 10^6 to 10^7 CFU/ml. A 1% inoculum, therefore, is sufficient to achieve population densities typical of surface waters ($\sim 10^5$ CFU/ml) and still keep background carbon levels and endogenous metabolism low. This is key for maintaining analytical sensitivity when degradation is assayed by nonspecific methods. In addition, rates of biodegradation exhibit relatively little variation when activated sludge inocula are compared with natural microbial populations. This is shown for glucose and two commercial surfactants in Table 6. Rate constants for biodegradation of these materials using an activated sludge inoculum agree well with rate constants for biodegradation observed in river water assays. These results indicate that sewage microorganisms can be used in the laboratory to predict rates of biodegradation in dilute natural systems. Extrapolation of laboratory rate data is also subject to fewer caveats since mixed cultures from an environmentally relevant system are being used.

Temperature

Within a physiological range, rates of biological reactions tend to be proportional to temperature. This proportionality is lost when temperatures exceed a maximum level and denaturation of cellular components occurs, or

TABLE 5. *Total number of CFU per milliliter in activated sludge supernatant after homogenization*

Expt	Date	CFU/ml[a]	Avg
SCAS I	July 1977	6.1×10^6 5.2×10^6 6.0×10^6	5.8×10^6
SCAS II	September 1977	1.3×10^6 1.5×10^6 1.2×10^6	1.3×10^6
SCAS III	October 1977	2.2×10^7 2.3×10^7	2.3×10^7
SCAS IV	December 1977	1.2×10^7 7.5×10^6 9.5×10^6	9.7×10^6
SCAS V	February 1978	1.3×10^7 1.5×10^7	1.4×10^7
SCAS VI	February 1978	2.0×10^7	

[a] Assayed in quadruplicate on nutrient agar.

TABLE 6. *Effect of sewage and river water inocula on rate constants for ultimate biodegradation*

Inoculum	Compound[a]	b (day^{-1})
1% Activated sludge	Neodol 45-7	0.13
	Neodol 23-6.5	0.10
	ATAC	0.25
	Glucose	0.26
Ohio River water	Neodol 45-7	0.12
	$C_{12}E_9$	0.09
	ATAC	0.36
	Glucose	0.21

[a] ATAC, Alkyl trimethyl ammonium chloride.

drop below a minimum level where enzyme control systems are inhibited. Between the maximum and minimum temperatures, the temperature dependency of biological reaction rates can often be empirically described by the Arrhenius equation:

$$y = Ae^{-E/RT} \tag{14}$$

where y is the reaction rate, E is activation energy, R is the gas constant, and T is absolute temperature.

The variation of rate constant with temperature can be written:

$$\frac{d \ln k}{dT} = \frac{E}{RT^2} \tag{15}$$

Integration with respect to T gives:

$$\ln k = \ln A - \frac{E}{RT} \tag{16}$$

A plot of $\ln k$ versus $1/T$, therefore, results in a straight line with slope = $-E/R$. This is usually how the temperature dependency of reaction rates is graphically determined.

The effect of temperature on rate constants for biodegradation also can be described by an Arrhenius relationship. As shown in Fig. 8, the rate constants for degradation of NTA and linear alkylbenzene sulfonate (LAS) are proportional to temperature, and degradation occurs at both high and low extremes. The effect of temperature on degradation is accurately described by the Arrhenius equation, with calculated activation energies of 14.5 and 9.1 kcal/mol for NTA and LAS, respectively. These activation energies are in the range observed for many enzyme-catalyzed reactions. Since degradation experiments are generally measured over a fairly narrow temperature span, linear plots of k versus temperature can also be used to predict rate constants at different temperatures (inset, Fig. 8).

It should be mentioned that rate constant estimates based on Arrhenius or linear relationships are likely to be conservative since they do not take into account the capabilities of specific psychrophilic or mesophilic populations. Microorganisms adapted to a particular temperature may degrade a material

much more efficiently than predicted on the basis of thermodynamic considerations alone. This is aptly illustrated by NTA monitoring studies conducted
in Canada, where NTA is used in detergent formulations. No significant increases were observed in the level of NTA in different surface waters during
winter or summer months, even though NTA usage remained fairly constant
and temperatures dropped to 0°C (17). In the real world, therefore, decreases
in degradation due to temperature were balanced by increases in efficiency
due to population.

Biodegradation rate constants are significantly affected by temperature, and
the Arrhenius equation is useful in predicting rate constants at different temperatures. In a broader sense, however, the real value of equation 16 lies in
its ability to determine whether the temperature dependency of a particular
degradation reaction is typical for biological systems and to put quantitative
bounds on this dependency. This process is key to removing subjective value

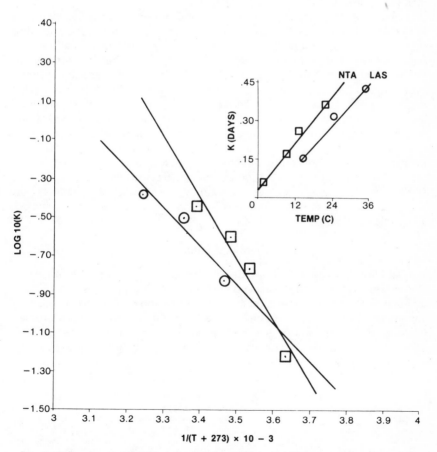

FIG. 8. *Arrhenius plots for biodegradation of NTA and LAS in environmental water samples. The linear approximation is shown in the inset.*

judgements about the degradation of a particular chemical and to placing observed results into a relevant biological perspective.

Dissolved Oxygen

Aerobic microorganisms have a very high affinity for molecular oxygen. Oxygen does not seriously affect the rate of bacterial metabolism until very low dissolved oxygen levels are attained. For example, the maximum rate for O_2 uptake by *Escherichia coli* (uMax) is not reduced until dissolved oxygen concentrations drop below 0.26 mg/liter (2). Since the rate of O_2 uptake is a saturable function of O_2 concentration as described by equation 1, a 90% approximation of the K_s for O_2 uptake is 10-fold below this concentration, or 0.026 µg/liter. This dissolved oxygen level is low enough that degradation of a chemical is likely to be limited by other factors (substrate concentration, temperature) before O_2 becomes rate limiting. The K_s values for O_2 uptake can be determined, however, if low dissolved oxygen levels are measured and maintained for extended periods of time.

Degradation of chemicals in environmental systems has been observed at low dissolved oxygen levels. Tiedje and Mason (11) found that NTA degradation occurred in soil at dissolved oxygen levels as low as 0.04 mg/liter, although at reduced rates when compared to saturation conditions. Tabatabai and Bremner (9) saw no effect of dissolved oxygen on the rate of NTA degradation in similar experiments. Degradation of NTA also occurs in surface waters (Table 7) at low dissolved oxygen levels, albeit at reduced rates compared to saturation levels. Based on the first-order rate constants, however, the half-life for NTA at low dissolved oxygen levels is still relatively short (<6 days). Thus, although O_2 levels in parts per billion range can decrease the rates of biodegradation, half-lives may not be significantly affected. In practical terms, dissolved oxygen levels are not likely to be a crucial variable in the majority of biodegradation studies unless they drop below 1 mg/liter. Verification of degradation under aerobic or microaerophilic conditions is much easier than under "presumed" anaerobic conditions.

Other Variables

In addition to temperature and dissolved oxygen, other variables must usually be regulated to ensure balanced microbial metabolism. These include regulation of pH and ionic strength, as well as provision of an adequate nutrient supply. Control of these variables in the laboratory is usually not a problem when synthetic media are used. However, the use of environmental water samples can present special problems. For example, highly buffered river water

TABLE 7. *Effect of dissolved oxygen (DO) concentration on rate constants for biodegradation of NTA*

DO (mg/liter)	k (day^{-1})	$t_{1/2}$ (days)
0.3–0.5	0.12	5.8
Saturation	0.53	1.3

or soil samples are efficient sinks for CO_2 evolved during mineralization studies, making the use of external CO_2 traps ineffective. Oligotrophic lakes or marine waters can also be limited for specific nutrients which restrict degradation of a chemical above a certain level. Nutrient limitation is a particular problem in estuarine systems (Escambia Bay, Fla.) where nitrogen/phosphorus limitations retard even glucose degradation above the 1 mg/liter level (R. D. Vashon, Procter & Gamble, personal communication). In general, conditions must be provided to ensure that the concentration of the test material is truly rate limiting for biodegradation and that the assay system is optimized to measure the rate of biodegradation.

SUMMARY

To briefly summarize the previous two sections, two major points can be stressed:

1. Rate constants for biodegradation can be determined in the laboratory for a variety of compounds by use of relatively simple batch systems and simple differential equations.
2. The effect of important variables can be incorporated into biodegradation rate expressions to predict the rate of biodegradation under a variety of conditions.

Therefore, it is possible to extrapolate biodegradation rate data from the laboratory to the environment and to compare rates of biological reactions with the rates of physical and chemical reactions. The combined rate data can then be used to predict environmental concentrations and, in combination with aquatic effect levels, can be used in the assessment of environment hazard.

ACKNOWLEDGMENTS

The excellent statistical assistance of R. L. Perry is gratefully acknowledged, as are the contributions of G. G. Clinckemaillie. Thanks must also be extended to R. D. Vashon and to other Environmental Safety Department staff for stimulating and thought-provoking discussions.

LITERATURE CITED

1. **Baughman, G.L., and R.R. Lassiter.** 1978. Prediction of environmental pollutant concentration, p. 35–54. In J. Cairns, Jr., K.L. Dickson, and A.W. Maki (ed.), Estimating the hazard of chemical substances to aquatic life. ASTM Special Technical Publ. 657. American Society for Testing and Materials, Philadelphia.
2. **Brown, D.E.** 1970. Aeration in the submerged culture of microorganisms, p. 125–174. In J.R. Norris and D.W. Ribbons (ed.), Methods in microbiology, vol. 2. Academic Press, London.
3. **Lighthart, B., and G.A. Loew.** 1972. Identification key for bacteria clusters from an activated sludge plant. J. Water Pollut. Control Fed. **44:**2078–2085.
4. **Lighthart, B., and R.T. Oglesby.** 1969. Bacteriology of an activated sludge wastewater treatment plant—a guide to methodology. J. Water Pollut. Control Fed. **41:**R267–R281.
5. **Pirt, S.J.** 1975. Principles of microbe and cell cultivation, p. 33, 67, 72. Halstead Press, New York.
6. **Smith, J.H., W.R. Mabey, N. Bohonos, B.R. Holt, S.S. Lee, T.W. Chou, D.C. Bomberger, and T. Mill.** 1978. Environmental pathways of selected chemicals in freshwater systems. Part II. Laboratory studies. Report No. EPA-600/7-78. U.S. Environmental Protection Agency, Athens, Ga.
7. **Soap and Detergent Association.** 1965. A procedure and standards for the determination of the biodegradability of alkyl benzene sulfonate and linear alkylate sulfonate. J. Am. Oil Chem. Soc. **42:**3–16.

8. **Stanley, P.M., and J.T. Staley.** 1977. Acetate uptake by aquatic bacterial communities measured by audioradiography and filterable radioactivity. Limnol. Oceanogr. **22:**26–37.
9. **Tabatabai, M.A., and J.M. Bremner.** 1974. Decomposition of nitrilotriacetate (NTA) in soils. Soil Biol. Biochem. **7:**103–106.
10. **Thompson, J.E., and J.R. Duthie.** 1968. The biodegradability of NTA. J. Water Pollut. Control Fed. **40:**306–319.
11. **Tiedje, J.M., and B.B. Mason.** 1974. Biodegradation of nitrilotriacetate (NTA) in soils. Soil Sci. Soc. Am. Proc. **38:**278–283.
12. **Wright, R.T.** 1973. Some difficulties in using ^{14}C-organic solutes to measure heterotrophic bacterial activity, p. 199–217. *In* L.H. Stevenson and R.R. Colwell (ed.), Estuarine microbial ecology. University of South Carolina Press, Columbia.
13. **Wright, R.T.** 1978. Measurement and significance of specific activity in the heterotrophic bacteria of natural waters. Appl. Environ. Microbiol. **36:**297–305.
14. **R.T. Wright, and B.K. Burnison.** 1977. Heterotrophic activity measured with radiolabelled organic substrates, p. 140–155. *In* Native aquatic bacteria; enumeration, activity and ecology. Proceedings, American Society for Testing and Materials Symposium, 25 June 1977, Minneapolis.
15. **Wright, R.T., and J.E. Hobbie.** 1965. The uptake of organic solutes in lake water. Limnol. Oceanogr. **10:**22–28.
16. **Wright, R.T., and J.E. Hobbie.** 1966. Use of glucose and acetate by bacteria and algae in aquatic ecosystems. Ecology **47:**447–464.
17. **Woodiwiss, C.R., R.D. Walker, and F.A. Brownridge.** 1979. Concentrations of nitrilotriacetate and certain metals in Canadian wastewater and streams: 1971–1975. Water Res. **13:**599–612.

Modeling the Environmental Fate of Synthetic Organic Chemicals

LENORE S. CLESCERI

Biology Department, Rensselaer Polytechnic Institute, Troy, New York 12181

The use of computer-programmed mathematical models can be an effective tool for the prediction of ultimate fate and effects of man-made chemicals in the environment. The conceptual framework for a modeling approach is detailed in this paper. The modeling approach is categorized into three separate areas: (i) environmental factors (pH, dissolved oxygen, temperature, and other physicochemical variables); (ii) biota and metabolic activity; and (iii) individual chemical structure. A description is provided of the model, PEST, developed for the prediction of an organic chemical's distribution within a lake ecosystem. The model embodies representations of all important physical, chemical, and biological processes that affect the transport and degradation of synthetic organic chemicals in aquatic environments. It can be used to model several segments of an aquatic system simultaneously with five temporal resolutions.

ENVIRONMENTAL EXTRAPOLATION OF DATA

In considering the usefulness of kinetic models in predicting the environmental fate of synthetic organic compounds, we should keep in mind certain general features about the natural environment: (i) it is a complex multivariable system, (ii) functional or even regulatory components may be imperfectly measured or even unidentified, and (iii) there is a strongly interconnected functioning of system components resulting in a reasonable overall "homeostasis." These features are strong arguments for the application of computer program modeling. Computer program models can be developed to simulate the complexity of ecosystems through: (i) delineation of the major components and interactions of the system, (ii) development of equations for the interactions to test conceptual framework, and (iii) assignment of numerical values for coefficients in these equations. Given this procedure, the question is, How good are these models and what is the accuracy with which they can be used to predict biotransformation and the fate of chemicals in the aquatic environment?

Models are developed at the process level ideally through close interaction between experimentation and procedure of mathematical representation of the process. This is the weakest point in most model construction. Usually, the data requirements are so great that a modeler will seek out any and all forms of data for development of the model. However, with the careful accumulation of data from literature, laboratory, and field investigation, models are developed which: (i) systematically organize diverse data, (ii) can be used to detect conceptual weaknesses in system function and point out areas of needed research (13), (iii) can be used to test difficultly tested hypotheses, and (iv) may or may not have predictive capability.

MODELING OF ENVIRONMENTAL FATE OF SYNTHETIC ORGANIC CHEMICALS

Depending on the particular system and the physicochemical properties of the chemical, the fate of a chemical in the aquatic environment may be (i) alteration or degradation by chemical, photochemical, or biological processes; short- to long-term storage in the sediment, water column, surface film, or biological tissue; or (iii) physical removal from the system via volatilization, dilution, or harvesting.

Many processes and many sites are involved in delineating the fate of chemicals in the aquatic environment. It is therefore rather obvious that a wide range of factors would be instrumental in affecting both rates of the processes and residence sites in the system. These are (i) environmental factors (pH, dissolved oxygen, temperature, sunlight, nutrients, mixing, particulate matter), (ii) biota and their metabolic activity, i.e., as modified by (i), and (iii) chemical structure. The processes that affect the chemical transformation of organic chemicals are depicted in Fig. 1. Those that affect storage are shown in Fig. 2.

The distribution of organic chemicals in a lake is illustrated in Fig. 3, which is a compartmental diagram of the model (PEST) developed by Rensselaer Polytechnic Institute's Center for Ecological Modeling. The particular processes involved in transport and biotransformation are shown in Fig. 4. These processes are assigned to the various compartments in Fig. 5.

DATA NEEDS: ENVIRONMENT

The model PEST embodies representations of all the important physical, chemical, and biological processes that affect the transport and degradation of synthetic organic chemicals in aquatic environments. It can be used to model several segments of an aquatic system simultaneously with five temporal resolutions.

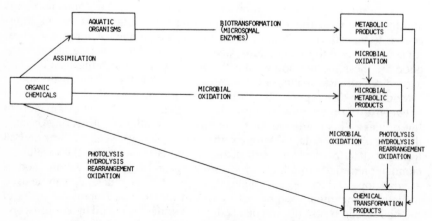

FIG. 1. *Environmental transformations of organic chemicals.*

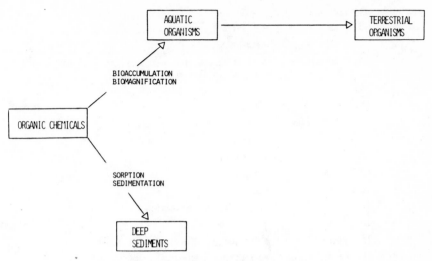

FIG. 2. *Environmental storage of organic chemicals.*

The processes resulting in chemical transformations are given in general by:

$$DGRG = HYD + OX + PHOT + MMET + BTRANS$$

where HYD is the rate of transformation due to chemical hydrolysis, OX is the rate of transformation due to chemical oxidation, PHOT is the rate of transformation due to direct and sensitized photolysis, MMET is the rate of trans-

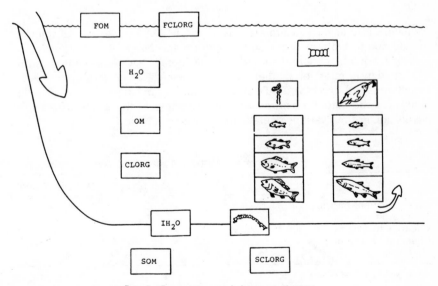

FIG. 3. *Compartmental diagram of PEST.*

PROCESSES

1. ADSRRP   ADSORPTION

2. DSRP   DESORPTION

3. BSRP, GSRP   BIOLOGICAL SORPTION - TO BODY (PASSIVE) OR TO
                  GILLS (ACTIVE)

4. VOL   VOLATILIZATION

5. C   CONSUMPTION

6. DEFS, EXCR   WASTE - DEFECATION AND EXCRETION

7. CRM   CREAMING

8. SED   SEDIMENTATION

9. RESUSP   RESUSPENSION

10. SOL   SOLUTION

11. BTRANS   BIOLOGICAL TRANSFORMATION

12. AGGR   AGGREGATION

13. DIFF   DIFFUSION

14. MORT   MORTALITY

15. BURIAL   DEEP BURIAL

16. MMET   MICROBIAL METABOLISM

17. CDRGD   CHEMICAL (AND/OR PHOTOCHEMICAL) DEGRADATION

18. OUT   OUTFLOW

19. IN   INFLOW

20. FLOAT   FLOTATION

21. GRND   GROUNDWATER MOVEMENT

22. DIL   DILUTION

FIG. 4. *Processes identified in PEST.*

formation due to microbial metabolism, and BTRANS is the rate of transformation due to biotransformation by other organisms.

From laboratory experiments and from the literature, one can compile arrays of degradation rates for synthetic organic chemicals. We have done this for malathion, methoxychlor, 2,4-D, atrazine, and carbaryl as part of our modeling study. For the natural environment, an inspection of these rates reveals the significance of the various degradation processes (Table 1).

TABLE 1. *Significant environmental degradation processes*

Degradation process	Malathion	Methoxychlor	2,4-D	Atrazine	Carbaryl
Biodegradation (microbial or microsomal)	×	×	×	×	×
Chemical hydrolysis	×	×	×	×	×
Chemical oxidation	—	—	—	—	—
Photolysis[a]	× S	× S	× D		× D

[a] S, Sensitized; D, direct.

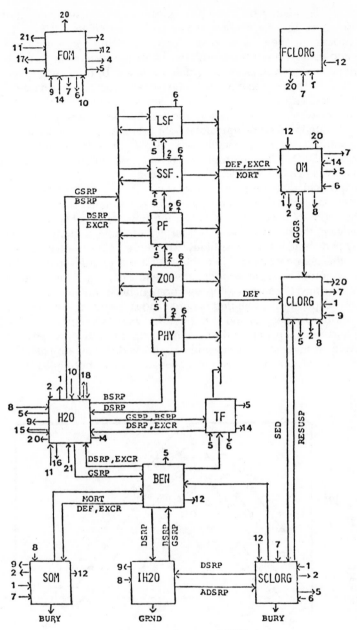

FIG. 5. *Flux diagram of PEST.*

Rates of degradation by these processes are, of course, dependent upon other variables. This can be best illustrated by some figures (Fig. 6–8).

The process of chemical decomposition of malathion in water is via elimination and carboxylate ester hydrolysis. As can be seen in Fig. 6, this process is accelerated at pH values above and below 4, and by increasing temperature in pure water. The degradation of methoxychlor proceeds by a base-catalyzed hydrolysis and HCl elimination. Figure 7 shows the effect of pH on the half-life of methoxychlor in pure water, and Fig. 8 shows the effect of pH on the photolysis rate induced by ultraviolet light (254 nm, mercury discharge lamp) in pure water. Longer wavelengths of sunlight will be less effective. The end products of 2,4-D (2,4-dichlorophenoxyacetic acid) photolysis have been reported as monochloro and dichloro derivatives.

It is apparent that the relative importance of the competing degradation processes depends on the particular set of environmental conditions of the receiving body of water. At this point, the importance of a rather complete knowledge of the receiving water is clear. For example, photolytic degradation of 2,4-D is dependent upon turbidity. Therefore, it is essential to predict accurately the seasonal changes in turbidity of a lake in order to predict the fate of 2,4-D. Another example demonstrates the role of oxygen for biodegradation of aromatic rings; the surrounding environment must contain dissolved oxygen (5). Thus, sorption to suspended material and subsequent sedimentation precludes further biodegradation if the benthic zone is anaerobic. Subsequent mixing and resuspension produces another opportunity, however, if the associated microflora is facultative and has survived the anaerobiosis.

Simple models are desirable and obtainable when the regulatory features

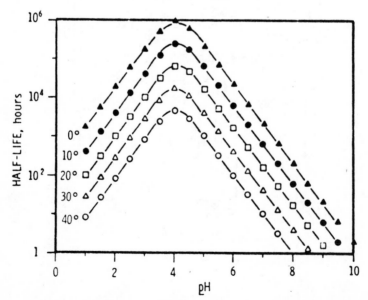

FIG. 6. *pH and temperature effect on malathion degradation (11).*

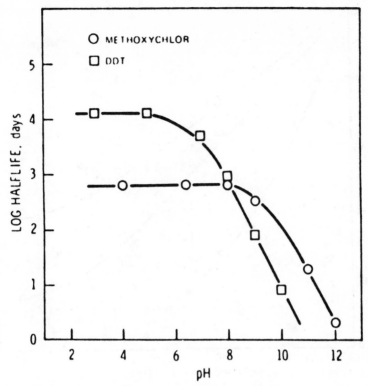

FIG. 7. pH and half-life profile for methoxychlor and DDT (dichlorodiphenyltrichloroethane) in water at 27°C (16).

of the system are known. The development of a protocol for the identification
of regulatory features is needed.

For some lakes, fertility can be predicted by phosphorus loading (15). Macroscopic modeling at Lake George has clearly identified phytoplankton primary
production, zooplankton predation of phytoplankton, and zooplankton nutrient excretion as the key processes in the cycling of phosphorus in Lake
George (12). Microbial biomass depends on the level of assimilable organic
matter in several Russian lakes and reservoirs (8). Microbial activity correlates
with the organic content of the oligotrophic lake sediments during the growing
season (2). These cause-and-effect relationships can be derived through testing
the sensitivity of a validated mathematical model to changes in compartmental
level and flux between compartments. This sensitivity analysis permits trim-
ming a model to the important processes. The key is the validation of the
model with field-collected data.

PROCESS DATA NEEDS

Each process contributing to synthetic organic degradation needs certain
site-specific data for model implementation. In PEST, for microbial metabolism,

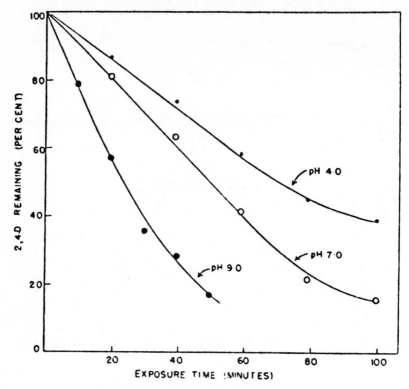

FIG. 8. *Effect of pH on photolysis in water (1).*

a maximum metabolic rate is measured in the laboratory and applied to a specific site. Application to the site requires measurements of dissolved oxygen, pH, temperature, mixing conditions, and total microbial biomass.

Microbial Metabolism Formulation

Microbial metabolism is defined as any chemical conversion of the parent compound by a microbial assemblage. It is assumed that microbial metabolism can occur wherever organic matter resides. PEST models microbial metabolism as

$$MMET = METMAX \cdot ACT \cdot EFFB \cdot [TOM/K_s + TOM]$$

METMAX = chemically and photochemically corrected TOM transformation rate by an "adapted" mixed assemblage under nonlimiting conditions of H^+, dissolved oxygen, temperature, non-energy-yielding nutrients, and mixing (days^{-1})

TOM = concentration of toxic organic matter (grams per liter)

K_s = constant equal to TOM at $\frac{1}{2}$ METMAX (grams per liter)

ACT = activity coefficient which converts METMAX to site conditions (unitless) and equals MIN (DOCOR, pHCOR, TCOR, MIXCOR)

EFFB = effective TOM degrading biomass (grams per liter) which equals
B · ADPOT where B = microbial biomass of the site (grams per liter) and
ADPOT = the adaptive potential of the microbial population (unitless) where

$$ADPOT = EXPT/ADPT$$

and where

EXPT = the time of exposure of the microbial biomass to TOM (days)
ADPT = the time it takes the microbes to develop a capability for degrading
the TOM under site conditions (days) and equals

$$MMGT \cdot ACT \cdot STRU$$

where

MMGT = generation time under METMAX conditions (days)
ACT = activity coefficient as above (unitless)
STRU = structural factor (unitless)

TERMS

Maximum Microbial Metabolism (METMAX)

In order to model microbial metabolism of TOM, a maximum value must be
determined which can subsequently be reduced by site correction factors (ACT
and EFFB) as specified in the above equation and discussed below. This maximum value ideally is the rate of degradation of the chemical by a microbial
assemblage with a high species diversity. This assemblage should have been
exposed to the compound for at least several generation times and should be
growing under optimal conditions of dissolved oxygen, temperature, pH, nutrients, and mixing. During this time some enrichment may also occur.

To provide for a wide range of biochemical activities, inocula from sources
which are undergoing complex organic decomposition should be combined
and used for the determination of METMAX, e.g., soil, marsh water, sediment.

In practice, the flocculant layer of a lake sediment exhibits a great deal of
flexibility by virtue of its interfacial position. It is subject to random fluctuations
in nutrients and dissolved oxygen as a function of mixing. It has rather constant or gradually changing temperature environs. As such, it is specialized for
many intermittent environmental conditions. This system amended with soil
and marsh inocula can be used for TOM degradation studies. Samples obtained
at peak seasonal temperatures for the system should be used since this condition would produce the most metabolically active assemblage.

Activity Coefficient (ACT)

The need to relate biodegradation capability to the site conditions is met
through the ACT term. This term modifies the maximum metabolic rate of the
effective biomass (EFFB), or the biomass capable of TOM degradation, to the
rate possible under site conditions. Thus, it is a reducing term determined by
the examination of site conditions and particular TOM. The environmental
parameter that is most restrictive to microbial growth and decomposition of
the particular TOM determines ACT. For example, if the TOM is aromatic and

the particular compartment being modeled is the sediment during summer conditions, the limiting parameter would be dissolved organic carbon, and a correction factor based upon the influence of dissolved oxygen on the metabolic rate of aromatic ring-degrading organisms would be used (DOCOR) to limit METMAX.

Effective Biomass (EFFB)

The term EFFB allows for the development of a TOM biodegradation capability through the process of adaptation. Adaptation may be genetic and occur through mutation or the acquisition of new genetic elements (transmissible plasmids). It may also be simply the time required for induction of suitable enzymes within the existing population. The term is derived through the use of an adaptation potential which expresses whether or not the adaptation will occur during the time of exposure to TOM. If the adaptation time exceeds the exposure time, the adaptation potential is set equal to 0, whereas if the adaptation time (ADPT) is equal to or less than the exposure time (EXPT), the adaptation potential is 1 or some fraction of 1 permitting a variation in the % TOM degraders in the total microbial biomass. Thus, the EFFB is determined by the measured microbial biomass in the system and the adaptive capability of the system.

Adaptation Time (ADPT)

The determination of ADPT can be made for a specific TOM along with the METMAX measurement. It can also be calculated for related compounds by using the ACT term, the generation time of the assemblage under METMAX conditions (MMGT), and a structural factor (STRU).

Mutation rate is dependent on generation rate under most conditions, but observed to be independent of generation rate under amino acid- or nitrogen-limited conditions in continuous culture (7). In addition to the mutation rate, the ability to develop a biodegradation capability for a certain TOM depends on the structural factor (STRU). It is currently estimated by means of the Hammett equation (6), which is used to predict relative biochemical reactivity from the nature of substituent functional groups of chemical derivatives.

The Hammett equation is

$$\log k/k_0 = \sigma p$$

where k_0 is the rate for the unsubstituted compound and k is the rate for the substituted compound.

However, the availability of reaction constant (p) and substitute constant (σ) values for aqueous reactions at biochemically realistic temperatures is limited.

With regard to ADPT, other approaches are being developed which follow the same rationale utilized by Enslein and Craig (4) in their toxicity estimation model for compounds culled from the Toxic Substances List. Their model was developed for estimation of acute toxicity based upon substructural fragment keys, partition coefficients, and molecular weights.

Another model utilizing general metabolic reactions and specific chemical reactions has been described by Spann et al. (14). This model addresses the problem of activated intermediates (e.g., malaoxon from the microsomal ox-idation of malathion).

These structure-activity studies developed for mammalian systems have a great deal of applicability to microbiological transformations. The concept that biological activity can be predicted from chemical structure is very old (3). Expanding the distribution prediction based on partition coefficients (9) to include transformation prediction based upon substructural fragments seems feasible for ecosystems as well as animal systems.

The usefulness of the calculated approach may be for deciding whether or not a particular TOM should be developed by an industry. Coupled with MET-MAX and MMGT for a related compound and ACT for the site which would be impacted by the new product, a reasonable estimate of environmental effect can be made.

Correction Factors (DOCOR, pHCOR, TCOR, MIXCOR)

To determine the value of ACT, various factors that affect ACT have to be examined to decide upon the limiting factor for a particular site and TOM. To determine the limiting factor, the site conditions of dissolved oxygen, pH, temperature, and degree of mixing for the sediment-water interface must be known. Some judgement is required to determine which factors are important for the particular TOM with respect to the degree of degradation. Based upon this judgement, that factor which is most limiting to the activity of the micro-organisms is selected to reduce ACT and subsequently convert METMAX to site conditions.

FEATURES OF THE MODEL

The microbial metabolism subprocess model of PEST is a dynamic repre-sentation of biodegradation and adaptation as influenced by environmental parameters and structure of TOM. It is different from other such models in that:

1. TOM degradation rate is a function of the activity of a mixed assemblage of microorganisms.
2. The composition of the assemblage can vary in % TOM degraders.
3. The effect of environmental parameters on the activity of the microflora is explicit.
4. The model can be utilized to develop structure-activity coefficients.

Relation to Other Approaches

We have discussed the approach of process rates and computer program modeling of the individual processes as one that allows the utilization of di-verse process data for the purpose of determining the overall effect of these processes on the fate of chemicals in the environment. An alternative, of course, is to utilize a real or model environment and measure the concentration of the chemical in the various phases of the test environment (10). This mi-crocosm approach has less flexibility than the mathematical modeling approach since each experiment is an integration of all of the individually defined processes affecting the fate of the chemical. In addition, since the experiment is run under a necessarily limited number of environmental conditions, the data are highly "site specific." It does, however, have the tremendous advantage of realism; i.e., there is the direct opportunity to observe the fate and effects of the chemical added to the system. It also can be used as a valuable tool in the validation of a mathematical model.

CONCLUSION

Our solution to the ever-increasing presence of toxic chemicals in the environment must be one of careful use, diligent monitoring, and utilization of all research tools that provide insight into the transport and transformations of synthetic chemicals in the environment. The use of computer-programmed mathematical models can be effective for many diverse applications when developed carefully, tested frequently, and used wisely.

LITERATURE CITED

1. **Aly, O. M., and S. D. Faust.** 1964. Studies on the fate of 2,4-D and ester derivatives in natural surface waters. J. Agric. Food Chem. **12:**541–546.
2. **Clesceri, L. S., and M. Dazé.** 1975. Relations between microbial heterotrophic activity, organics and deoxyribonucleic acid in oligotrophic lake sediments. Verh. Int. Ver. Limnol. **19:**974–981.
3. **Crum-Brown, A., and T. Fraser.** 1868–1869. Trans. R. Soc. Edinburgh **25:**151, 693.
4. **Enslein, K., and P. N. Craig.** 1978. A toxicity estimation model J. Environ. Pathol. Toxicol. **2:**115–121.
5. **Gibson, D. T.** 1972. Initial reactions in the degradation of aromatic hydrocarbons, p. 116–136. *In* Degradation of synthetic organic molecules in the biosphere. National Academy of Sciences, Washington, D.C.
6. **Jaffé, N. N.** 1953. A reexamination of the Hammett question. Chem. Rev. **53:**191–261.
7. **Kubitschek, H. E., and H. E. Bendigketi.** 1961. Latent mutants in chemostats. Genetics **46:**105–122.
8. **Kusnetsov, S. I.** 1970. The microflora of lakes and its geochemical activity (translated and edited by C. H. Oppenheimer). University of Texas Press, Austin.
9. **Leo, A. C., Hansch, and D. Elkins.** 1971. Partition coefficients and their uses. Chem. Rev. **71:**525–616.
10. **Metcalf, R. L., and P. Y. Lu.** 1973. Environmental distribution and metabolic fate of key industrial pollutants and pesticides in a model ecosystem. Report U1LU-WRC-0069, University of Illinois Water Resources Center, Champaign.
11. **Paris, D. F., D. L. Lewis, and N. L. Wolfe.** 1975. Rates of degradation of malathion by bacteria isolated from aquatic system. Environ. Sci. Technol. **9:**135–138.
12. **Roberts, G.** 1979. Data based macroscopic modeling at Lake George. FWI Newsletter **9**(2):1–4. Rensselaer Polytechnic Institute's Fresh Water Institute, Troy, N.Y.
13. **Scavia, D.** 1979. The use of ecological models of lakes in synthesizing available information and identifying research needs, p. 109–168. *In* Perspectives on lake ecosystem modeling, Ann Arbor Science Publishers, Inc., Ann Arbor, Mich.
14. **Spann, M. L., D. C. Chu, W. T. Wipke, and G. Ouchi.** 1978. Use of computerized methods to predict metabolic pathways and metabolites. J. Environ. Pathol. Toxicol. **2:**123–131.
15. **Vollenweider, R. A.** 1969. Possibilities and limits of elementary models concerning the budget of substances in lakes. Arch. Hydrobiol. **66:**1–36.
16. **Wolfe, N. L., R. G. Zepp, J. A. Gordon, G. L. Baughman, and D. M. Clines.** 1977. Kinetics of chemical degradation of malathion in water. Environ. Sci. Technol. **11:**88–93.

Synopsis of Discussion Session: Extrapolation

P. H. PRITCHARD, R. J. LARSON, AND L. S. CLESCERI

The discussion resulting from the session on the extrapolation of laboratory data to natural environmental situations dealt with three major problem areas: (i) the steps and decision points involved in providing definitive exposure concentration estimates, (ii) the experimental limitations to biodegradation rate constants, and (iii) the assessment of the effects of environmental parameters on degradation rates and their integration into predictive mathematical models. All of these problems indicated a need for more research before definitive extrapolations can be made. It appears, however, that our approaches to the problem of extrapolation are headed in the right direction and that the results coming forth from various research laboratories lead to a certain degree of optimism. As discussed in the papers presented in this session and others, the potential of reducing biodegradation rates to simple descriptive rate constants which may be applicable over a large range of environmental variables is indeed exciting. Certainly, the ensuing discussions indicated that the rate constant approach was a provocative concept which should greatly stimulate and orient a data-gathering effort. This effort could eventually culminate in an appreciation of biodegradation rates which would substantially increase our confidence in making exposure concentration predictions through the extrapolation of laboratory data to the environment.

The steps and decision points involved in providing a definitive estimated exposure concentration (EEC) are potentially numerous and tedious. With the time constraints before us, the laboratory and field investigations of these steps, although ultimately necessary, may greatly slow parts of our regulatory process. The demands, therefore, for the establishment of an EEC procedure which will generate results with good confidence intervals must be weighed carefully. Exposure concentration estimations can, in fact, be generated with various degrees of sophistication or confidence intervals, or both. For example, to couple, at this point in time, definitive EEC values (or its methodological requirements) for individual compartments within an aquatic ecosystem with a less definitive toxicological data set for the development of an environmental safety factor is inappropriate; in this case, it is overkill with the EEC values. As outlined in Fig. 1, both definitive environmental effect assessment and definitive exposure concentration estimation are processes which will require time and more research. In both situations, however, early-on or range-finding assessment and estimations are readily possible, and they can potentially be evaluated together to develop safety factors for chemicals and regulatory decisions. It cannot be denied that regulatory decisions in the past have been based upon empirical laboratory toxicology data for which the relationship to *actual* environmental effects has never been established. It is thus possible at the present time to enhance these decisions on toxicological effects with laboratory biodegradation data whose relationship to the actual fate of chemicals in the environment has not yet been fully established. Accordingly, our discussions have indicated that the confidence in laboratory biodegradation and

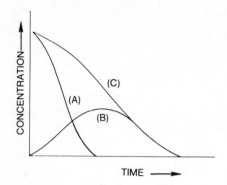

FIG. 1. *Hazard evaluation process for new and expanded-use chemicals showing the interdependence of effect data and fate data. Analysis of the biodegradation of an alcohol ethoxylate shows the effect of analytical method on information given. (A) Specific nonionic analysis: wickbold shows primary biodegradation. (B) Thin-layer chromatography: identifies polyethylene glycol residues. (C) Nonspecific analysis: dissolved organic carbon, shows ultimate biodegradation.*

fate data may be approaching that which is presently in existence for laboratory toxicological data, and thus many reasonable decisions in planning and regulation can be made, now, for a large base of chemical types.

As we push toward the development of definitive exposure concentration evaluations through the extrapolation of laboratory data to the real world, we must keep in mind that it is a process which will continually test and verify (i.e., reduce our confidence intervals) decisions we have made previously; it will *not*, in fact, generate a magical all-encompassing environmental concentration number for the entire aquatic system.

Exposure concentration estimations, therefore, have two levels:

1. *Range-finding estimation (rough cut)*—This estimation is obtainable with present data and existing methodology. Acceptable and unacceptable limits, based historically on past biodegradability tests of various chemical classes, do not now exist, but will have to be formulated in the near future. The range-finding estimation follows a working scheme basically like that outlined in the earlier paper by C. M. Lee. It can be integrated with other fate process measurements by using simple material balance models extrapolated over a hypothetical lake, river, or pond such as those developed in this section by Branson and Blau.

2. *Definitive estimations (refined)*—These estimations will require time to develop data sets and the testing of concepts. Environmental mathematical models of generalized lakes, rivers, and estuaries such as described by Clesceri will be employed. These estimations will eventually modify range-finding estimations.

RATE CONSTANTS

To assess the influence of biodegradation processes on exposure concentration estimations, at either level, a quantitative evaluation of the rates at which these processes occur is necessary. Since measurements of reaction rate constants tend to be highly specific for individual and system-independent measurements, degradation rate constants are needed in order to extrapolate laboratory data to the environment. Accordingly, for the purpose of determining the effects of environmental parameters on biodegradation rate constants, which is felt to be a major step in extrapolation, it is necessary to establish experimental conditions in the laboratory where the rate of decrease in the concentration of a substrate, ds/dt, is pseudo-first order with respect to the concentration of that substrate $[S]$:

$$-ds/dt = k[S] \tag{1}$$

where k is the pseudo-first-order rate constant.

Substrate disappearance may be represented by loss of toxic properties of a chemical (primary degradation) or by changes in metabolic parameters (oxygen consumption, carbon dioxide evolution) as long as the equivalence of different measurements is established. Rate constants should be determined in dilute systems by exposing acclimated microbial populations to multiple concentrations of the substrate under conditions where $[S]$ is low (0.01 estimated k_s) and rate limiting. Low concentrations allow k to be determined in the absence of saturation effects for nongrowth substrates. For growth substrates, rate constants may be affected by biomass; it is important to keep $[S]$ low so that biomass changes do not affect k. When $dB/dt \ll dS/dt$, dS/dt is still a first-order function of $[S]$ and $dS/dt = k[S]$.

Extrapolation of rate constants between different dilute aquatic systems will depend heavily on the relative importance of biomass concentrations $[B]$ in determining biodegradation rates. First-order rate constants, k, for mineralization of test materials which support microbial growth appear to be independent of biomass when it is in the range of 10^3 to 10^6 ml. On the other hand, rate constants for primary degradation of nongrowth substances must be normalized against $[B]$ in order to extrapolate results between different dilute systems. These apparent differences may be due to the growth or nongrowth nature of compounds tested or to the different methods of biomass measurement. Extensive research will, therefore, be required to determine whether dS/dt is (i) independent of biomass at levels typically found in surface waters, (ii) directly proportional (first order) to biomass levels, or (iii) related to biomass in a manner dependent on our ability to accurately determine the relative amount of active biomass in a particular system.

If rate constants for biodegradation cannot be extrapolated to environmental systems on the basis of the first two relationships above, then significant research effort will be required under the third to determine the effects of biomass on k. This effort may include both development of methods to determine the percentage of the microbial population active on a certain compound or the activity of this population based upon suitable kinetic parameters.

EFFECTS OF ENVIRONMENTAL PARAMETERS

The validity of extrapolating laboratory-derived data to biodegradation in the various compartments of the natural environment depends upon several factors:

1. The true substrate (toxic organic substances) dependence of the biodegradation pseudo-first-order rate constant

2. The ability to describe the biological, chemical, and physical characteristics of recipient environments in a manner which will quantitatively assess and predict the extent to which they will modify the biodegradation (as well as photochemical and chemical degradation) rate constants for each specific compartment. This will include the level of toxic organic substance in the compartment.
3. The establishment of equivalency in microbial biomass measurements employed in the rate determinations and in the accurate characterization of the environmental compartment to which the rate constant is being applied
4. The tendency of the microflora in the recipient compartment to develop a biodegradation capability (acclimation) in a manner similar to the laboratory condition
5. Reliance upon carefully developed environmental fate models to relate toxic organic substance inputs into the aquatic environment with distribution and concomitant transformation within the various aquatic compartments. This procedure is based upon the development of differential equations that describe the dominant carriers and toxic organic substrate fluxes between carriers as functions of the particular biological, physical, and chemical conditions of the aquatic compartment.

From an extrapolation standpoint, the effect of environmental parameters on degradation rate constants was considered to be the most critical problem in the development of a definitive EEC. It is reasonable and quite probable that spatial and temporal differences in some environmental parameters will not be great enough to produce a significant change in the biodegradation rate constant because of the diverse capabilities of natural populations of microorganisms. However, this needs to be tested further. The environmental parameters listed below are considered as important factors that could affect the laboratory-derived biodegradation rate constants:

* Surface area
* pH
* Mixing
  Organic and inorganic nutrients
* Temperature
* Salinity
  Redox potential
  Dissolved oxygen
  Microbial diversity

The factors marked by asterisks are probably predictable or will be predictable after a relatively short series of experimental studies. At some point, effects will have to be assessed through some functionally designed experiments. Those unpredictable factors which do affect the rate constant will need to be quantified for each new chemical or series of related chemicals tested. If the rate constant is insensitive to changes within the expected environmental range, then the factor is considered unimportant for that chemical. If there is a dependence, the importance of that dependence on the EEC can be determined through sensitivity analysis with the mathematical model. Caution should also be exercised in evaluating this dependence relative to the range of experimental error associated with the rate constant determination; if experimental errors are large, they erroneously dilute the effects of certain environmental factors.

EECs in various environmental compartments as predicted by suitable fate models for lakes, rivers, and estuaries will be a specific value taking into account the compartment dynamics of the particular ecosystem. The practical application of the EEC values will depend in part on the ability of regulatory agencies to adequately describe or delineate the environmental compartment that is most critical to their hazard assessment program.

VALIDATION

Environmental exposure estimations generated from predictive mathematical models will have limited physical meaning until suitably validated. This is a factor relative to development of confidence in these predictions. Several options exist for the validation step.

1. Monitoring or field studies can be implemented. The most useful information will come from field sites which are being directly dosed either intentionally or unintentionally. In some instances, specific field enclosures can be studied.

2. The environmental distribution of a number of conservative organic chemicals (DDT [dichlorodiphenyltrichloroethane], polychlorinated biphenyls, Kepone, phthalates, etc.) has been well documented. The physical and chemical properties of these materials (including rate constants) are also readily available. It is therefore possible to feed information to a predictive model for these compounds and then compare its predicted distribution and fate with its actual historical record in the environment under consideration. At the very least, parameters such as sediment adsorption, transport and mixing, bioaccumulation into food chains, etc., can be estimated and calibrated from available data for the generalized river, lake, or estuarine conditions.

3. Aquatic environments are often difficult systems in which to perform field studies, if only because they have high dilution capacity and they are at the mercy of climatic and seasonal fluctuations. If parts of the environment under study can be physically brought into the laboratory and studied under a standard set of conditions, then some common appreciation of environmental structure and function can be ascertained. This is commonly referred to as microcosm studies, and we believe that they have excellent potential as validation tools for exposure concentration estimates. We define a microcosm as a calibrated laboratory simulation of a portion of a natural aquatic environment in which environmental components, in as undisturbed a condition as possible, are containerized under a standard set of laboratory conditions. By calibrated, we mean two things. First, the simulation *must* be checked by comparing common properties of the microcosm (functional processes, for example) with similar properties in the environment. And, second, the conditions in the microcosm or the fate results it generates *must* be checked for their relation to the system design.

For example, we stated that if an acclimation in rate of degradation occurs at low substrate concentration in a laboratory biodegradation test system, then it is likely to occur in the environment. This would be difficult to test in the field. However, a microcosm, as a surrogate to some part of the environmental situation, could serve as an excellent tool for assessing whether acclimation actually occurs in nature. Its demonstration in a microcosm would say that it occurs in a system which is the next best thing to a field study. As long as specific questions are asked of a microcosm, it would appear to be a promising validation tool.

4. Finally, it is important to keep in mind that the validation process could go on ad nauseum with no real gain in accuracy of the EEC prediction. At some point, it may be necessary to arbitrarily draw a line, stop testing, and consider the validation as state of the art. Subsequent regulatory decisions will then have to be contested before more testing may be required.

Chapter 6. QUANTITATIVE EXPRESSION OF BIOTRANSFORMATION

Quantitative Expression of Biotransformation Rate

GEORGE L. BAUGHMAN, DORIS F. PARIS, AND WILLIAM C. STEEN

Environmental Research Laboratory, U.S. Environmental Protection Agency, Athens, Georgia 30613

Methodologies and approaches to the study of microbial transformation of organic compounds in natural waters and sediments are outlined. Tentative evidence is given to demonstrate that use of rate constants is a reasonable approach. Application of second-order kinetics incorporating the effects of organic compound concentration, bacterial concentration, and sorption by suspended sediment is made to natural systems.

It is now generally conceded that hazard assessment requires some understanding of the fate of chemicals in the aquatic ecosystem. This requirement stems from the need to forecast the exposure concentration in order to predict impact. The problems and approaches to hazard assessment were the subject of another meeting at Pellston, Mich., in 1977 (6). At that time, it was pointed out (3, 5) that environmental concentrations of a pollutant could be predicted if the rates of important processes controlling pollutant concentration can be determined. This is easily summarized by equation 1.

$$\text{rate}_{net} = \text{rate}_{in} - \text{rate}_{export} - \text{rate}_{abiotic} - \text{rate}_{biological} \tag{1}$$

Here, the terms represent the net rate of change in pollutant concentrations, rates of change due to input, export and volatilization, abiotic transformation, and biological transformation, respectively. It is not our purpose here to examine this hypothesis and its associated assumptions but rather to discuss the feasibility of adequately describing the final term in equation 1, i.e., the kinetics of microbiological transformation.

DISCUSSION

Microbial kinetics is not a new area of research, having been used for many years in fermentation processes, drug manufacture, and waste treatment (1, 2). In most cases, however, the approaches have relied on the empirical equation first suggested by Monod (10) (equation 2) or modification thereof.

$$\frac{d[B]}{dt}\left(\frac{1}{[B]}\right) = u = \frac{u_m[S]}{K_s + [S]} \tag{2}$$

where $[B]$ is microbial population or activity per unit volume, $[S]$ is substrate concentration, u_m and u are maximum specific and specific growth rates, respectively, and $K_s = [S]$ when $u = \frac{1}{2}u_m$.

Because so much microbial kinetic work has been done using this equation, a brief assessment of its suitability for environmental purposes is in order. This equation has usually been modified (equation 3) by addition of a yield coefficient ($Y = d[B]/d[S]$) which simply describes the efficiency with which substrate is converted to biomass because substrate transformation rate is required rather than growth rate.

$$\frac{d[B]}{dt}\left(\frac{d[S]}{d[B]}\right) = -\frac{d[S]}{dt} = \frac{u_m[B][S]}{Y(K_s + [S])} \tag{3}$$

The practical difficulty of basing kinetics on measured values of Y is not obvious except in a system having a single limiting substrate. Under these conditions, Y must approach zero as $[S]$ approaches zero because of maintenance metabolism. In nature, however, the pollutant is not the sole substrate, and so the proper value or values of Y are not readily determined from laboratory studies.

Because the condition of chief concern is one of long-term, low-level input to the environment, the concentration of substrate can be assumed to be much less than K_s. This does not seem too unreasonable since values for K_s typically range from 0.1 to 10 mg/liter, which is much higher than the concentration of most pollutants. Under this circumstance, equation 3 can be approximated by equation 4, which is in the familiar form of second-order kinetics.

$$-\frac{d[S]}{dt} = \frac{u_m}{YK_s}[B][S] = k[B][S] \tag{4}$$

Equivalence of u_m/YK_s and k has been demonstrated experimentally for systems where the substrate was the sole carbon source and second-order kinetics were observed (11, 12).

The rationale for use of a second-order rate expression derived from the Monod equations is thus tentative, but it is certainly well known that microbial rates are a function of population size and substrate concentration at higher concentrations. The possibility of a threshold at low concentrations has been speculated on by Boethling and Alexander (4), but the data are inconclusive. These equations also neglect the problem of acclimation or induction and describe the rate only after such processes have occurred. This does not, however, seem inconsistent with the need to predict rates that might occur in a system receiving input over a long period of time.

Tentative acceptance of a second-order equation as the appropriate phenomenological description leaves a number of questions of a conceptual, and thus experimental, nature. How is $[B]$ to be defined and measured? How is the substrate defined and measured, i.e., CO_2 equivalence, specific compound analysis, chemical oxygen demand, etc.? Assuming that the kinetics are second order and that the rate constant can be measured for a natural system, what is the variability in the rate constant from one ecosystem to another? These questions will now be addressed in reverse order.

Table 1 shows second-order rate constants that were measured (D. F. Paris, W. C. Steen, and G. L. Baughman, Natl. Meet. Am. Chem. Soc., 175th, Anaheim, Calif., March 1978) for a number of different waters. The results were obtained by the river die-away method, except that plate counts were used to determine population size, and substrate concentration was maintained low enough so that the kinetics were pseudo-first order in substrate. Examination

TABLE 1. *Second-order degradation rate constants for three compounds in water*

Compound	k (1 organism^{-1} h^{-1})[a]	N[b]	Temp (°C)[c]
2,4-D[d] butoxyethyl ester	$(5.4 \pm 0.50) \times 10^{-10}$	31	18.8 + 6.6 (1–29)
Malathion	$(4.4 + 0.66) \times 10^{-11}$	14	21.1 ± 7.8 (2–27)
Chlorpropham	$(2.4 + 0.42) \times 10^{-14}$	11	19.1 ± 5.6 (7–28)

[a] Mean across all sites and all concentrations ± relative standard deviation.
[b] Number of sites for which k was determined.
[c] Average temperature ± standard deviation. Value in parentheses is the range of temperature.
[d] 2,4–D, 2,4–Dichlorophenoxyacetic acid.

of the mean and standard deviation (Table 1) across all sites clearly shows that the mean value is highly independent of site. Studies at different concentrations of substrate and organisms confirmed that the rate is first order both in substrate and in population size for all cases examined. This surprisingly high level of predictability raises many questions which can only be answered by many more data. It does offer significant encouragement, however, for prediction of microbial transformation kinetics.

TABLE 2. *Rate constants for transformation in sediment-water systems*

Compound	Fraction sorbed	k_{app} (1 organism^{-1} h^{-1})	k_2 (1 organism^{-1} h^{-1})
2,4-D butoxyethyl ester[a]	0[b]	$(2.3 \pm 0.8) \times 10^{-10}$	
Malathion[a]	0[b]	$(4.0 \pm 0.9) \times 10^{-11}$	
Chlorpropham[c]	0[b]	$(1.42 \pm 0.6) \times 10^{-13}$	
	0.15	$(4.15 \pm 0.7) \times 10^{-13}$	$(3.45 \pm 0.2) \times 10^{-13}$
	0.90	$(7.4 \pm 1.2) \times 10^{-15}$	$(2.29 \pm 0.6) \times 10^{-13}$
Methoxychlor[a]	0[b]	$(6.7 \pm 1.2) \times 10^{-14}$	
	0.15	$(5.2 \pm 2.3) \times 10^{-14}$	$(3.0 \pm 0.5) \times 10^{-14}$
	0.90	$(6.1 \pm 2.8) \times 10^{-16}$	$(3.2 \pm 1.2) \times 10^{-14}$
Bis(2-ethylhexyl) phthalate[a]	0[b]	$(4.2 \pm 0.7) \times 10^{-15}$	
	0.15	$(1.8 \pm 0.3) \times 10^{-15}$	$(3.6 \pm 1.3) \times 10^{-15}$
	0.90	$(5.7 \pm 1.7) \times 10^{-17}$	$(6.1 \pm 2.2) \times 10^{-15}$
Dibutyl phthalate[c]	0[b]	$(3.1 \pm 0.8) \times 10^{-11}$	
	0.15	$(2.7 \pm 1.5) \times 10^{-11}$	$(3.5 \pm 1.7) \times 10^{-11}$
	0.90	$(6.1 \pm 0.5) \times 10^{-13}$	$(2.6 \pm 0.2) \times 10^{-11}$
Di(n-octyl) phthalate[a]	0[b]	$(3.7 \pm 0.6) \times 10^{-13}$	
	0.15	$(2.1 \pm 1.2) \times 10^{-13}$	$(3.6 \pm 0.7) \times 10^{-13}$
	0.90	$(3.1 \pm 0.4) \times 10^{-15}$	$(6.9 \pm 2.3) \times 10^{-13}$

[a] Data from W. C. Steen (in preparation). 2,4-D, 2,4-Dichlorophenoxyacetic acid.
[b] Sediment free.
[c] Data from Steen et al. (13).

These measurements were made by following the disappearance of the parent compound, which in effect defines transformation. Gilbert and Watson (7) call this primary or functional biodegradation, i.e., biodegradation of a substance to an extent sufficient to remove some characteristic property of the molecule. It should be noted that the rate of primary transformation will probably not be equal to the rate of mineralization or loss of absorbance, etc. For preliminary assessment of environmental behavior, the rate of primary degradation will usually be the parameter of choice. For other purposes such as treatability, mineralization rate may well be required.

The question of how the active population, [B], should be measured is controversial. Jannasch and Jones (8), Kangas and Coleman (9), and others have compared different methods. It is unlikely that this question has a single best answer. The only real requirement is that the method be adequately reproducible and comparable from one site or study to another. The data in Table 1 indicate that, at least for these compounds, plate counts are adequate.

For degradation studies in suspended, aerobic, sediment systems, plate counts cannot be used, and it was found to be necessary to resort to adenosine triphosphate measurement (13). Table 2 shows the rate constants obtained via equation 5:

$$-\frac{d[S]_T}{dt} = \frac{k}{1 + \rho K_p} [B][S]_T = k_{app}[B][S]_T \qquad (5)$$

where k is the second-order rate constant from equation 4, ρ is the sediment-

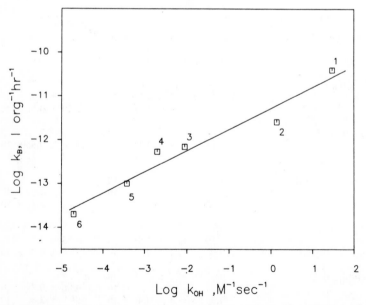

FIG. 1. Correlation of second-order alkaline hydrolysis rate constants (27°C) determined in distilled water using second-order biolysis rate constants determined in natural water samples. The compounds are (1) n-butoxylethyl ester of 2,4-D (2,4-dichlorophenoxyacetic acid), (2) malathion, (3) methyl benzoate, (4) methyl anisate, (5) methoxychlor, (6) chlorpropham.

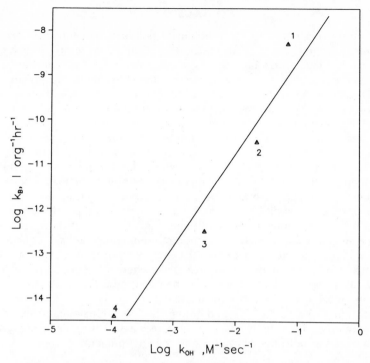

FIG. 2. *Correlation of second-order alkaline hydrolysis rate constants (30°C) obtained in distilled water with second-order biolysis rate constants (25°C). (1) Dimethyl phthalate, (2) di(n-butyl) phthalate, (3) di(n-octyl) phthalate; (4) di(2-ethylhexyl) phthalate.*

to-water mass ratio, K_p is the partition coefficient of substrate between sediment and water, k_{app} is the apparent second-order rate constant, and $[S]_T$ is the total amount of substrate per unit volume of the system.

These data indicate that, when population size and partitioning are included, the process is again first order in $[B]$ and $[S]$. Also, the data for the compounds in Table 1 show the second-order rate constants to be about the same whether measured with water column organisms or with sediment populations in the absence of sediment. Further, the influence of sediments on degradation of highly insoluble (hydrophobic) compounds is clearly expected to be a lowered rate, all other things being equal.

If rate constants that are system independent can be measured, then it may be possible to develop relationships between the rate constants and properties that reflect molecular structure. The use of linear free energy (structure-activity) relationships has been well known for many years in chemistry and more recently in biology. Figures 1 and 2 show the relationships between the second-order biolysis rate constant and the abiotic second-order alkaline hydrolysis rate constant (14). The good correlation of these data over so many orders of magnitude lends considerable credibility and promise to the potential utility of this approach. Why such a good correlation exists is not clear at this time, but the explanation may well be because the microbial transformation is hydrolytic for these compounds.

CONCLUSION

Application of kinetics to the problem of pollutant transformation in the environment requires that we formulate the equations in terms of substrate disappearance rather than growth. Also, the problem must be considered in the context of multiple carbon sources; i.e., the substrate is not growth limiting. Data have been presented to show that a simple second-order rate equation can be used and that, for the study compounds, the rate constant is highly independent of the system for which it was measured. These observations leave unanswered many questions about the role of temperature, cometabolism, other nutrients, population structure, and the structure of the compound. There is little compelling evidence in the literature to suggest that a similar approach cannot be taken with most compounds. On the other hand, confidence in the approach and the ultimate answer to these questions can only be obtained by additional kinetic studies for many compounds and sites. Perhaps most importantly, future studies must incorporate statistical treatment of data and mathematical analysis according to some proposed rate equations. Unless this is done, no basis exists for comparison of results. After all, the most important question for our purpose is not whether the rate be predicted; rather, it is necessary to know how well the rate can be predicted in order to determine whether the results are adequate.

Finally, kinetic models as outlined here provide a rationale that is essential for development of methodologies to rigorously assess the role of microbial transformation in both the laboratory and the environment.

LITERATURE CITED

1. **Aiba, S., A. E. Humphrey, and N. F. Millis.** 1973. Biochemical engineering, 2nd ed. Academic Press, London.
2. **Bailey, J. E., and D. F. Ollis.** 1977. Biochemical engineering fundamentals. McGraw-Hill Book Co., New York.
3. **Baughman, G. L., and R. R. Lassiter.** 1978. Prediction of environmental pollutant concentration, p. 35–54. In J. Cairns, Jr., K.L. Dickson, and A.W. Maki (ed.), Estimating the hazard of chemical substances to aquatic life. ASTM Special Technical Publ. 657. American Society for Testing and Materials, Philadelphia.
4. **Boethling, R. S., and M. Alexander.** 1979. Effect of concentration of organic chemicals on their biodegradation by natural microbial communities. Appl. Environ. Microbiol. **37:**1211–1216.
5. **Branson, D. R.** 1978. Predicting the fate of chemicals in the aquatic environment from laboratory data, p. 55–70. In J. Cairns, Jr., K.L. Dickson, and A.W. Maki (ed.), Estimating the hazard of chemical substances to aquatic life. ASTM Special Technical Publ. 657. American Society for Testing and Materials, Philadelphia.
6. **Cairns, J., Jr., K. L. Dickson, and A. W. Maki (ed.).** 1978. Estimating the hazard of chemical substances to aquatic life. ASTM Special Technical Publ. 657. American Society for Testing and Materials, Philadelphia.
7. **Gilbert, P. A., and G. K. Watson.** 1977. Biodegradability testing and its relevance to environmental acceptability. Tenside Deterg. **14**(4):171–177.
8. **Jannasch, H. W., and G. E. Jones.** 1959. Bacterial populations in sea water as determined by different methods of enumeration. Limnol. Oceanogr. **14:**128–139.
9. **Kangas, M. J., and D. C. Coleman.** 1975. Compound of five techniques for the estimation of numbers and activity of bacteria in soil. Bull. Ga. Acad. Sci. **28:**143–148.
10. **Monod, J.** 1949. The growth of bacterial cultures. Annu. Rev. Microbiol. **3:**37–394.
11. **Paris, D. F., D. L. Lewis, J. T. Barnett, and G. L. Baughman.** 1975. Microbial degradation and accumulation of pesticide in aquatic systems. Report No. EPA-660/3-75-007. U.S. Environmental Protection Agency, Athens, Ga.
12. **Smith, J. H., W. A. Mabey, N. Bohonos, B. R. Holt, S. S. Lee, T.-W. Chiu, D. C. Bomberger, and T. Mill.** 1978. Environmental pathways of selected chemicals in freshwater systems. Part II. Laboratory studies. Report No. EPA/600-7-78-074. U.S. Environmental Protection Agency, Athens, Ga.

13. **Steen, W. C., D. F. Paris, and G. L. Baughman.** 1979. Effects of sorption of toxics to natural sediments on microbial degradation. *In* Proceedings of the Symposium, Processes Involving Contaminants and Sediments, 177th National Meeting of the American Chemical Society, Honolulu, April 1979 (in press).
14. **Wolfe, N. L., D. F. Paris, W. C. Steen, and G. L. Baughman.** 1980. Correlation of microbial degradation rates with chemical structure, in press.

Synopsis of Discussion Session: Quantitative Expression of Biotransformation

H. L. BERGMAN, G. L. BAUGHMAN, D. F. PARIS, AND W. C. STEEN

Forecasts of expected environmental concentrations (EEC) require quantitative biotransformation data. Although it is recognized that nonquantitative methods may have their place in studies such as metabolite identification or pass-fail degradation tests, the use of mathematical system models to forecast the EEC in typical aquatic environments requires that the rates of all important transformations (e.g., biotransformation, photolysis, etc.) be described mathematically. Biotransformation rates are not suitable for EEC prediction because the rates are functions of conditions used in the kinetic determination. On the other hand, rate constants are presumably independent of environmental factors and reproducible from site to site. Thus, the mathematical formalism used to describe biotransformation should be a rate equation from which rate constants can be derived.

It has been shown here that the second-order rate constants for microbial hydrolysis of several different compounds were reasonably reproducible from site to site. The rate equation used was

$$d[S]/dt = k[B][S]$$

where $[B]$ is the microbial population or biomass, $[S]$ is the substrate concentration (pollutant), and k is the second-order rate constant. This expression and others may prove suitable for describing many biotransformation reactions.

Whatever experimental procedures and mathematical expressions are used, however, several cautions should be observed with respect to measurement of biomass and substrate for rate equations. A reproducible method for estimating active biomass is essential for the determination of a rate constant and for predicting a rate based on known rate constants. No one method emerges as clearly superior in all cases. Comparisons are needed for biomass estimates obtained by various methods (plating techniques; acridine orange direct counts; measurement of cellular components such as adenosine triphosphate, deoxyribonucleic acid, or lipopolysaccharide; and activity measurements such as phosphatase, dehydrogenase, or heterotrophic uptake of some simple substrate). It remains to be seen whether total biomass will correlate with rates of transformations other than the hydrolytic pathway described in this chapter. It may be necessary to develop a means for estimating the numbers of organisms able to perform specific transformations.

When possible, substrate disappearance should be measured for rate constant determinations, since this is the term required in the kinetic computation and since substrate transformation rate is the process that must be used in estimating exposure concentrations. If analytical limitations dictate the use of other parameters such as oxygen utilization or carbon dioxide formation, then the functional relationship between substrate transformation rate and the rate of the measured process must be demonstrated for concentrations of interest.

Substrate concentrations used in biotransformation determinations should approximate expected environmental concentrations. At such low concentrations, mathematical analyses may be simplified if the kinetics are pseudo-first order in substrate. Another reason for keeping the test concentrations low is that high concentrations may permit microbial populations to adapt and acquire a transformation ability that would not exist under environmental conditions.

In addition to the above concerns about determination of biomass and substrate, several other questions must be addressed to improve our use of rate constants in forecasting expected exposure concentrations. For instance, the influence of important environmental parameters (O_2, pH, temperature, etc.) on rate constants has not been adequately determined in the laboratory or field. Further, rate constant data are available for only a few transformation pathways and a few sites. If rate constants vary substantially from site to site for some transformations, the variations should be addressed and accounted for. However, for the purposes of forecasting expected exposure concentrations from biotransformation rate constants, attention must be focused on similarities and on assessing the magnitude and importance of differences.

CONCLUSIONS

1. We are optimistic that reproducible rate constants can be determined for many if not all biotransformation reactions and that these rate constants can be used to forecast expected exposure concentrations for aquatic ecosystems.

2. We need to continue to improve our understanding, determination, and use of biotransformation rate constants, especially in relation to the measurement of microbial biomass, the effect of environmental parameters on rate constants, and the reproducibility of rate constants for different microbial transformations.

3. We need to continue to define the precision attainable in the determination of rate constants from different sites and to understand the significance of differences that are observed.

Chapter 7. FATE OF CHEMICALS IN THE AQUATIC ENVIRONMENT: CASE STUDIES

Nitrilotriacetate: Hindsight and Gunsight

JAMES M. TIEDJE

Departments of Crop and Soil Sciences and of Microbiology and Public Health,
Michigan State University, East Lansing, Michigan 48824

The environmental history of nitrilotriacetate, a readily biodegradable and highly water-soluble replacement builder for detergents, is discussed from both a scientific and a regulatory viewpoint. Nitrilotriacetate data are used as an example demonstrating the need for meaningful predictions of environmental concentrations, for reproducible test methods, and for good extrapolation of laboratory data to actually existing environmental conditions.

Nitrilotriacetate (NTA) is not one of your normal environmental pollutants; it is readily biodegradable, highly water soluble, and of very low toxicity. Why all the fuss? Perhaps NTA was the right chemical but at the wrong time. Let's recall the scenario. First, an environmentally conscious public finally enacted controls on phosphate pollution of waters, probably without recognition of the problem of what should be the replacement "builder" in detergents. NTA was the most acceptable candidate on the scene, but was not really in a position to be suddenly released in the huge tonnages required by this use. The environmental and toxicological data were limited, but this large introduction would not have been wise in any case. Second, the government had just given birth to the Environmental Protection Agency (EPA), and the public was looking for responsiveness from its new savior. Because of the timing of the events— creation of EPA and then the ban on NTA—it is hard for me to not feel that NTA was being used as evidence of action. (Major detergent manufacturers, in response to concerns of the surgeon general, voluntarily discontinued use of NTA in December 1970, pending further study. Although this initial decision was the surgeon general's, EPA has certainly been influential in the history of NTA.) Some plan of testing and gradual expansion of use was clearly needed, but the action was in effect a ban. It created an uncertainty—do you drop it or do you invest in it—which affected industry, university researchers, and government agencies. Also, as a result of the ban, NTA became a household word; detergents were proudly advertised as "contains no NTA." Even the then Secretary of the Interior appeared in television ads promoting supposedly wholesome detergents. The marketability of NTA was destroyed. This would probably have been the end of NTA except for a third aspect. Canada chose the other road and has continued to use NTA to the present (now 15% of detergent content) (6). Thus, we had a new kind of experiment, Canada the treatment and the United States the control. Canada's results are positive, and thus the implication is that the United States could move to further reduce phosphates and replace them with NTA. But old decisions are hard to change.

This scenario has taken 10 years; extensive data now exist on NTA so that reliable decisions can be made, but who cares? The forces of the past have created a new stage, new characters, and probably a new plot. What is to be learned? Perhaps two things—first, science moves too slowly, decisions are made before 10 years, and second, when the issue is in the political-public arena, the ground rules are different, and objective science has not learned how to be an influential voice. Science is done by individuals who speak with individual voices. Furthermore, scientists are trained to be conservative interpreters of data and have difficulty dealing with decisions where 95% confidence intervals are not known.

Recently, NTA has undergone an extensive review on both the ecological and health aspects by scientific committees appointed by the International Joint Commission of Canada and the United States. Their reports to the Research Advisory Board form a very current and thorough review and analysis of the situation (5, 6). The Ecological Task Force "concludes that the extensive literature and Canada's experience indicates that the use of NTA does not constitute an obvious environmental hazard" but does suggest nine specific areas that need more work. In my opinion, those nine areas are not critical but serve more to complete the entire story and thus would lend a greater degree of confidence to any decision.

Since this workshop has identified the areas of environmental concentrations, testing methods, extrapolation, and quantitative expression for focus, I will make a few comments from the NTA example that are important to these topics.

1. *Interrelation of environmental concentration and uptake kinetics.* Currently, this is one of the most interesting areas of both basic and practical microbial ecology. The question is, How do organisms utilize a compound present in nature at very low substrate concentrations, be it glucose or NTA? Because of the extreme oligotrophic nature of the open ocean, this has been a central question to marine microbial ecologists for some time. Their thought and research is leading the way, though they are finding it an experimentally challenging task. As a generalization, I would expect active transport to be necessary for significant metabolism at these low substrate concentrations. For the pollutant case this is troublesome since it is unlikely that specific permeases exist for these unusual anthropogenic substrates. For NTA, we believe that active transport and a permease exist. The cells have a low K_m, 18 μM (3.4 ppm), for metabolism of NTA (3), a respectably low value for an organism that would function in nature. Even with a low K_m, one would, of course, still expect to find a finite concentration in nature. This is exactly the Canadian experience; they found an average of 5 ppb of NTA in their natural waters, with 96% of the samples containing less than 25 ppb (6). It would seem useful to determine the relationship between input NTA, density of the NTA-degrading community, kinetic parameters of the NTA-degrading organism, and output NTA concentration. This model should be useful in predicting response to NTA input fluctuations, in predicting NTA steady-state output levels, and in maintaining an NTA-degrading population. In particular, the relationship between kinetic parameters and steady-state output concentrations might be a useful parameter for predicting concentrations of chemicals in any natural aqueous system. It is sometimes forgotten that enzyme kinetics dictate that there will always be some finite concentration of a chemical—even glucose—in an ecosystem.

Using NTA as an example, I have calculated rates of NTA degradation for natural waters to show why measurable concentrations will always remain. I have used the V_{max} (0.21 μmol of O_2 consumed per min per 4.3 mg of cells) and K_m (18 μM) data we found for an NTA-degrading pseudomonad (3). Since the concentrations found in nature are well below the K_m, I have reduced the Michaelis-Menten equation to the simple first-order case where $V_{max}/K_m = k$, the first-order rate constant, and the expression becomes $V = k$ [S]. I have also assumed that 1 mg of cell mass equals 10^9 cells. Using these values, the second-order k is 6.03×10^{-8} ml · min^{-1} · $cell^{-1}$.

In Table 1, I have used this value to calculate rates of NTA degradation at different concentrations of NTA for different temperatures and densities of NTA degraders. The estimate of NTA degrader density of 10^4 cells per ml is an assumption based on these degraders constituting about 1% of the total bacterial population and is probably an upper limit.

Taking the 5-ppb value at 10°C, which is a reasonable case for the Canadian situation, the rate of 0.0060 ppb · h^{-1} means that it would take 7 days for the concentration of 5 ppb to be reduced to 4 ppb, providing there are no new inputs. Hence, it should be apparent why NTA exists in Canadian waters.

In support of this explanation, it should also be noted that amino acid concentrations in natural waters range from about 5 to 100 ppb (2). Thus NTA concentrations can be considered typical of those other amino acids.

The above case for NTA with active uptake is probably unusual for most pollutants since many are hydrophobic. These should partition into the membrane and may remain isolated from the catabolic enzymes of the cell. Thus, I would expect grazing of microbial cells to be an important fate of the pollutant and biomagnification to be the result of significance.

2. *Methods of study of biodegradation.* The fact that bacteria could be grown on NTA as sole carbon source was established early (cf. 8). Thus, one immediately felt more comfortable that a serious problem of permanent contamination would not likely occur. I believe this interpretation is important and should receive substantial weight. Likewise, establishing whether a compound supports growth or is metabolized by cometabolism should be an important objective. The velocity (V) versus substrate concentration ([S]) relationship should be very different in the two cases. For cometabolism I think finding a high K_m and low V, relative to more known metabolites, might be a good indicator of cometabolism as well as being useful for predicting kinetic behavior.

This relationship implies a low rate which is first order with respect to substrate to a very high [S]; in fact, saturation may never be observed. This is

TABLE 1. *Predicted rates of nitrilotriacetate (NTA) degradation in nature as a function of NTA concentration, density of NTA-degrading bacteria, and temperature*

NTA (ppb)	Predicted rates of NTA degradation (ppb·h^{-1})		
	10^4 cells/ml, 30°C	2.5×10^3 cells/ml, 20°C	10^3 cells/ml, 10°C
1,000	36.2	4.52	1.21
25	0.90	0.113	0.030
5	0.18	0.022	0.006
1	0.036	0.0045	0.0012

exactly the case we found for metabolism of another chelant, ethylenediaminetetraacetic acid, which we believe is cometabolized (7).

3. *Acclimation*. When the compound serves as a growth substrate, variability due to acclimation is more likely and thus is important to predicting behavior. Certainly, degradation in a system with 0, 1, and 10^4 organisms per liter is very different.

With NTA, I have always wondered what is the true substrate for these seemingly specific enzymes. Why should these enzymes be conserved during evolution unless there is a natural substrate? If we knew what the natural substrate was, we would have a better idea of acclimation and maintenance of populations capable of degradation.

4. *Methodology*. I believe that ^{14}C methodology is the only way that good biodegradation work can be done. With binding to particulate materials now recognized as an important fate, using a label is the only way this can be quantified and its permanence can be evaluated. In the case of NTA, other sensitive methods were not available at first, and even now the nonlabel methods are far more involved and expensive. With ^{14}C studies, questions such as purity, position of label, and products measured (e.g., only $^{14}CO_2$) become important in the interpretation.

5. *Extrapolation*. To me, this means good kinetics. The basic response can then be modified by temperature, pH, etc. Simple models should be helpful, manageable, and reliable—useful tools. This was never done for NTA to my knowledge, probably because biodegradation occurred so readily.

6. *Diversity*. Diversity of microorganisms degrading NTA was never examined. Are there many strains, a few, or only one that can degrade NTA? How widely are they distributed? Do they have the same pathway, the same kinetics? We compared the three best-studied organisms and found that the latter two seemed identical, yet one was isolated from Michigan and the other from British soil (4). I think knowing whether a basic diversity exists is important to having confidence in the general degradation potential as well as to explaining the accuracy (or inaccuracy) of any predictions.

The impacts of these characteristics and several others upon the experimental design and ultimate test interpretations are further discussed in Table 2 for the two major cases, one in which the chemical is a growth substrate and the other in which it is not. The latter is often referred to as cometabolism, a term which has met with mixed reception. I think it can be a useful term when used to describe the two basically different types of patterns of degradation, as I have done in Table 2. It is less useful when considered at the molecular level since it could be due to any of several phenomena (1) already well known by other terms. Most cases of cometabolism are probably due to nonspecificity of enzymes, a phenomenon known since the discovery of enzymes. Why shouldn't one expect a substrate to be metabolized by an enzyme previously recognized to metabolize another substrate?

Table 2 gives the characteristics and expected response of the two major patterns of biodegradation. The characteristics and kinetics can be used to identify which pattern is present. I believe the points of emphasis should be different for the two cases. For the growth case, more emphasis should be placed on acclimation, whereas in the cometabolism case, more emphasis should be placed on accurate measurement of rates (extrapolation). Overall,

TABLE 2. *Summary of the different characteristics and expected behavior of the two general types of biodegradation, one in which the chemical is a growth substrate and the other in which it is not*[a]

Topic	Growth substrate	Not a substrate for growth (cometabolism)
Characteristics	Organism will grow as sole C source. Generally ultimate degradation. No lag after second addition of substrate.	Organism will not grow as sole C source. Accumulation of intermediate products likely. No difference in pattern between first and later additions of substrates.
Kinetics	Exponential; increasing [E]. Low K_m, high k.	First order; constant [E]. High K_m, *generally low* k^b; saturation may not be seen.
Behavior at low [S]	Possible anomalous behavior due to threshold for induction.	No anomalous behavior; first-order kinetics apply throughout.
Acclimation	Major effect; lag may be quite variable or lengthy due to patchiness and low initial density of degraders, and perhaps starvation state of organism in natural sample.	Often no effect; rarely causes induction, may increase tolerance to toxic chemical.
Relation of degradation kinetics to total activity or biomass, e.g., [B] for second-order rate expression	Likely not valid, more likely to be habitat specific, and, because of low densities, patchy.	May be valid since activity of interest is often proportional to general biomass or activity.
Extrapolation	General: expect eventual degradation in nature. Quantitative: difficult to be precise because of growth kinetics and acclimation effects, but may not be important problem because of generally fast rates; environmental effectors of less significance.	*Measure kinetic parameters accurately:* because of the generally slower rates, extrapolations will be made over longer times, and thus measured parameters need to be accurate. Also environmental effectors, e.g., temperature, pH, play a more important role.
Effect of added carbon	Diauxie pattern.	Generally effect is proportional to population unless specific carbon source happens to induce or repress/inhibit activity of interest.
Transport into cell	Active transport more likely.	Passive transport, thus likely slow and perhaps rate limiting.
Decision making	Generally, NO PROBLEM.	Needs MORE TESTING; proportionally more effort needs to be devoted to chemicals in this class.

[a] These are intended to be generalizations to guide approaches and interpretations; exceptions are expected. Abbreviations: [E], concentration of rate-limiting enzyme; k, second-order rate constant; [S], substrate (i.e., test chemical) concentration; [B], expression for concentration of total heterotrophic biomass or activity, or consistent fraction thereof.

[b] Except for easily degraded head types, a.g., hydrolytic cleavages

more effort needs to be placed on chemicals that are cometabolized since those are much more likely to yield environmental problems.

LITERATURE CITED

1. **Dagley, S.** 1979. Summary of the conference, p. 540–541. *In* A. W. Bourquin and P. H. Pritchard (ed.), Microbial degradation of pollutants in the marine environment. Publ. EPA-600/9-79-012. U.S. Environmental Protection Agency, Washington, D.C.
2. **Fenchel, T., and T. H. Blackburn.** 1979. Bacteria and mineral cycling, p. 55. Academic Press, London.
3. **Firestone, M. K., and J. M. Tiedje.** 1975. Biodegradation of metal-nitrilotriacetate complexes by a *Pseudomonas* species: mechanism of reaction. Appl. Environ. Microbiol. **29:**758–764.
4. **Firestone, M. K., and J. M. Tiedje.** 1978. Pathway of degradation of nitrilotriacetate by a *Pseudomonas* species. Appl. Environ. Microbiol. **35:**955–961.
5. **International Joint Commission.** 1977. Health implications of NTA. International Joint Commission, Great Lakes Research Advisory Board, Windsor, Ont.
6. **International Joint Commission.** 1978. Ecological effects of nonphosphate detergent builders: final report on NTA. International Joint Commission, Great Lakes Research Advisory Board, Windsor, Ont.
7. **Tiedje, J. M.** 1977. Influence of environmental parameters on EDTA biodegradation in soils and sediments. J. Environ. Qual. **6:**21–26.
8. **Tiedje, J. M., B. B. Mason, C. B. Warren, and E. J. Malec.** 1973. Metabolism of nitrilotriacetate by cells of *Pseudomonas* species. Appl. Microbiol. **25:**811–818.

Role of Biodegradability in the Environmental Evaluation of Polychlorinated Biphenyls and Chemicals in General

V. ZITKO

Fisheries and Environmental Sciences, Department of Fisheries and Oceans, Biological Station, St. Andrews, New Brunswick E0G 2X0, Canada

A summary of the environmental history of polychlorinated biphenyls is presented, with particular emphasis on the questions of environmental persistence. It is shown that biodegradation tests played a relatively minor role in the detection of environmental problems associated with polychlorinated biphenyls. A discussion of the variable nature of biodegradability tests currently used throughout North America, Europe, and Japan is provided, leading to the suggestion that more attention should be given to the development of realistic, workable schemes conducted within a scientifically sound, efficient, and economical framework.

This paper presents a very brief history of polychlorinated biphenyls (PCBs) and of the detection of the "PCB problem." It illustrates that biodegradation tests played a relatively minor role during the discovery of this problem.

In addition, various proposed national and international requirements for the assessment of toxicity and environmental impact of chemicals are discussed. It is suggested that more attention should be given to practical applications of the proposed systems, to develop a realistic scheme that is workable, scientifically sound, efficient, and economical.

HISTORY OF PCBs

The first PCBs were produced in 1929 by the Swann Corporation for use as nonflammable heat-transfer media, and in 1930 their production in the United States was taken over by Monsanto Co. (13). By 1936, it became apparent that workers exposed to PCBs develop dermatitis, and by 1938, yellow atrophy of the liver was attributed to PCB exposure (10). Consequently, PCBs were recognized as occupationally hazardous chemicals, and threshold limit values for exposure of workers to PCBs in air were established.

Analytical methods for the measurement of PCBs in air were based on the so-called lamp combustion. PCBs were absorbed in a combustible organic solvent, usually amyl acetate, and burned; the vapors were passed through an aqueous sodium hydroxide; and the chloride ion was measured by titration with silver nitrate (10).

This method was not specific enough to detect the accumulation of PCBs in biological samples, and in any case, the fate of PCBs in the environment was not considered. With the advent of DDT (dichlorodiphenyltrichloroethane), it was realized some 10 years later that persistence in the environment was an important factor in the environmental impact of a chemical. Persistence was at first valued as an asset, until improved analytical techniques detected the accumulation of DDT in tissues and biomagnification of DDT in food chains became apparent. Even then, the attention was on pesticides and not on industrial chemicals, until PCBs were identified in samples of wildlife in 1966.

In retrospect, it appears that the lack of ability to measure PCBs in biological samples and to detect their accumulation in biota were the crucial factors leading to the development of the PCB problem. Good analytical techniques are the safeguards against the occurrence of another "PCB problem," since bioaccumulation warrants a very thorough toxicological evaluation and risk/benefit assessment.

For a number of years after the identification of PCBs in the environment, the main emphasis of research was on the documentation of the ubiquity of the contamination of the environment and on the levels of PCBs in fauna, primarily in fish, marine mammals, and birds. Studies on the toxicity of PCBs started at about the same time, and as the testing protocols became more and more refined, the toxic threshold concentrations were decreasing. Studies on sources and transport of PCBs in the environment, photochemical degradation, and metabolism started to appear later.

Results of studies on the microbial degradation of PCBs were reported relatively late in the emergence of the "PCB problem."

MICROBIAL DEGRADATION OF PCBs

Keil et al. (11) reported that Aroclor 1242 stimulated the growth of Escherichia coli. On the other hand, Bourquin and Cassidy (6) reported that growth of Flavobacterium, Bacillus, Corynebacterium, Pseudomonas, Achromobacter, Micrococcus, and Serratia, isolated from a Florida estuary, was inhibited by Aroclor 1242. At the same time, Wong and Kaiser (16) showed that Aroclors 1221, 1242, and 1254 did not inhibit the growth of bacteria isolated from lake water. In addition, about 1% of lake water bacteria could use Aroclor 1221 or 1242, but not Aroclor 1254, as sole source of carbon. The bacteria were species of Achromobacter and Pseudomonas.

The differences in the results of these studies indicate that PCBs are not particularly powerful inhibitors of bacterial growth and that the effects of PCBs may depend to a large extent on the experimental conditions. It appears that growth studies carried out in this manner are not very useful for the assessment of the biodegradability of a chemical.

In 1973, Ahmed and Focht (1) demonstrated the biodegradability of several mono- and dichlorobiphenyls by Achromobacter and identified chlorobenzoic acids as degradation products. Baxter et al. (4) used Nocardia and Pseudomonas species, known to degrade biphenyl, and found that dichlorobiphenyls were degraded relatively rapidly with a half-life of about 10 days. The half-life of trichlorobiphenyls was 20 to 40 days, and little if any degradation of tetrachlorobiphenyls was observed.

Tucker et al. (15) found that the rate of biodegradation of commercial Aroclors by semicontinuous activated sludge decreased practically linearly with increasing chlorine content:

$$R = 106 \ (\pm 6) \ - \ 1.7 \ (\pm 0.2)^* \ D$$

(R = percent degradation in a 48-h cycle; D = percent chlorine). The decrease of biodegradability with increasing content of chlorine was confirmed by Furukawa and Matsumura (7). An *Alcaligenes* species, isolated from a lake sediment and able to use biphenyl as a sole carbon source, was used. Tetrachlorobiphenyls were slightly degraded, and some degradation was observed even with 2,4,5,2',5'-pentachlorobiphenyl. Substituted α-hydroxymuconic acids were identified tentatively as the biodegradation products or intermediates. In a subsequent paper, Furukawa et al. (8) correlated structure and biodegradability of chlorobiphenyls by species of *Alcaligenes* and *Acinetobacter*. 2,2'-Dichlorobiphenyls and 2,6'-dichlorobiphenyls were degraded more slowly than other dichlorobiphenyls, and the same pattern was observed for trichlorobiphenyls. Relatively little degradation of tetra- and 2,4,5,2',5'-pentachlorobiphenyl was obtained. Average degradation rates are given in Table 1.

It seems that, had such biodegradation studies been performed before the commercial introduction of PCBs, and had the general awareness of environmental problems been in 1930 as it was in 1970, PCBs, with the possible exception of Aroclor 1221 and similar preparations containing only mono- and dichlorobiphenyls, would not have been introduced on a commercial scale, provided that the toxicity or bioaccumulation had been known at the same time.

The PCB problem was actually discovered when it became obvious from analytical chemistry data that PCBs are persistent and accumulate in biota; biodegradation studies played a relatively minor role.

EVALUATION OF BIODEGRADATION STUDIES

For PCBs, the biodegradation studies produced relatively clear-cut results. The rate of biodegradation decreases with increasing chlorine content of the preparation; it is very low for most tetrachlorobiphenyls; pentachlorobiphenyls and more chlorinated biphenyls are practically non-biodegradable.

It is easy to make a decision on the basis of a practically complete lack of biodegradation as was observed with highly chlorinated biphenyls. It is also easy to make a decision on the basis of a rapid biodegradation. The difficulties arise with compounds that are degraded at a medium to a relatively slow rate, as observed with trichloro- and tetrachlorobiphenyls, were one to rely on the biodegradation data only. Bioaccumulation and toxicity data will help to arrive at a decision, but again only if the data are clear-cut, that is, either high bioaccumulation or high toxicity or lack of both. Again, decision problems are likely to be encountered with compounds that are "mid-scale."

TABLE 1. *Primary biodegradation rate of chlorobiphenyls (8)*

Chlorobiphenyl	Rate (nmol/cell per h)
Mono	7×10^{-8}
Di- (both rings substituted)	6×10^{-8}
Tri- (both rings substituted)	5×10^{-8}
Tetra- (both rings substituted)	2.5×10^{-8}
Penta-[a] (both rings substituted)	1.5×10^{-8}

[a] Only one isomer, 2,4,5,2',5', studied.

ENVIRONMENTAL LEGISLATION

Environmental legislation is in effect in a number of countries, and data requirements and testing protocols are being developed. It may be of interest to compare the approaches taken in various countries on the testing and evaluation of biodegradability. Much of the documentation exists in the form of drafts at different stages of consideration, and I cannot be certain that the information available to me is up to date.

In general, the data requirements on chemicals appear to be quite similar and consist basically of nomenclature, physical and chemical properties, biodegradation data, and toxicity data. As one advances along this outline to more complex requirements, the differences between the requirements of different countries or international organizations tend to increase.

For example, in Great Britain the list is limited to a minimum of information that is needed to identify primary hazards by a panel of experts. It is at the discretion of this panel to indicate whether any or which additional tests are required. A comprehensive list of tests is avoided since such an approach is considered nonsensical, on both scientific and economic grounds (12).

In Canada, under the Environmental Contaminants Act, the government may request information on the manufacture or import of chemicals that the government suspects are entering or are likely to enter the environment and endanger either human health or the environment. At present, there are 17 chemicals on the list. The list is subdivided further into several categories, depending on the available evidence of harm. Chemicals in the highest category are banned or regulated. For example, PCBs, polybrominated biphenyls, polychlorinated triphenyls, and mirex are in this category. Chemicals manufactured or imported for the first time in quantities exceeding 500 kg per year must be reported. In addition, a list of properties similar to that required by Great Britain must be given.

In the United States, reporting guidelines as well as test protocols are being developed. The tests required are quite complex, and it seems to me that, just as Great Britain may use the lower limit of the testing requirements, the testing requirements of the United States represent the upper limit; other countries are somewhere between.

Of the international organizations, the Organization for Economic Cooperation and Development is developing an evaluation scheme for chemicals. The scheme is in a state of flux, and it is too early to comment on it.

With the exception of Japan, it seems that most countries concentrate on the development of complex evaluation schemes, and little or no attention is paid to the numerical results of the tests and some guidelines or scales by which these could be evaluated.

In Japan, where the law is based on the determination of biodegradability, bioaccumulation, and chronic toxicity, numerical guidelines for the test evaluation are in effect (14). If biochemical oxygen demand (BOD) (14 days) of a compound under standard conditions (activated sludge; ratio of compound to sludge, 3:1; concentration of the compound, 100 ppm) exceeds 30% of the theoretical oxygen demand, the compound is usually judged well biodegradable. For bioaccumulation, if a bioconcentration factor in carp after 8 weeks of continuous exposure exceeds several hundred, the compound is usually judged highly bioaccumulative. Chronic toxicity tests are based on "teratogenicity, tumorigenicity, and pharmacokinetics, etc." The evaluation criteria for reaching decisions from these tests are not given in the paper by Sasaki (14).

By using the Japanese criterion of biodegradability, for example, phthalates are biodegradable, whereas many nitrocompounds and, to indicate the sensitivity of the criterion, compounds such as octylphenol, p-tert-butylphenol, anthracene, and terphenyl are non-biodegradable. That the criterion may also fail, if applied indiscriminately, is demonstrated by listing lead stearate as a biodegradable compound.

Another evaluation scheme that includes a numerical rating of the test results was developed by Astill et al. (3). For biodegradability evaluation, the ratio of BOD5 to COD (chemical oxygen demand) is used. Numerical ratings of 1, 2, and 3 are given to compounds with BOD5/COD values of >0.5, 0.25 to 0.5, and <0.25, respectively.

WHERE DO WE GO FROM HERE?

In the case of PCBs, it was the toxicity to occupationally exposed workers that sounded the first alarm. The alarm was not listened to. Undoubtedly, in spite of all kinds of tests, we are likely to encounter such situations again, but this time, health effects in occupationally exposed workers would certainly trigger a comprehensive investigation and toxicity testing.

Relatively simple bioaccumulation and biodegradation tests would have detected undesirable properties of PCBs, and if nothing else, they would have at least prompted additional testing.

Good analytical techniques are, of course, a prerequisite for all tests. Had better analytical techniques been available in the 1930s, the potential for widespread environmental contamination by PCBs might have been detected. Our analytical techniques are much better now; they are perhaps too good in certain instances, but at the same time there are many compounds that cannot be measured reliably. With respect to some of these compounds, we may be in a situation corresponding to that of PCBs in the 1930s. High-molecular-weight chlorinated paraffins are an example of this. Chlorinated paraffins are produced in amounts approaching 250×10^6 kg annually on the world scale. There are no reliable routine analytical techniques for chlorinated paraffins, and no meaningful biodegradation tests have been carried out (9). Recent data show that some types of chlorinated paraffins are accumulated by fish (5).

It seems to me that the set of people drafting various complex schemes for the evaluation of chemicals is somewhat removed from first-hand bench experience, and the drafts thus contain some far-fetched, impractical, and even scientifically unfounded suggestions. Before we get too far in drafting such schemes, it would certainly be useful to keep checking their efficiency and practicality on cases of some well-known and studied chemicals. These chemicals could then also be used as bench marks for future assessments of other chemicals.

The Environmental Studies Board, Commission on Natural Resources, recently carried out such an assessment of PCBs. It may be a good idea to perform similar studies on chemicals such as DDT, parathion, leptophos, phthalates, phenol, and aniline to see how these would fare in the different evaluation schemes, and, if these chemicals do present a hazard to the environment, in which tests and in what sequence these hazards would have been indicated most efficiently.

Lists of substances approved by the Japanese environmental legislation are also available (see, for example, references 2 and 14) and may be a good source for comparing various test and evaluation schemes.

An efficient evaluation scheme that protects human health and the environment from the majority of possible risks and yet is reasonably simple, is not overly expensive, and does not hinder innovation must be the ultimate goal. We seem to be still quite far from it.

LITERATURE CITED

1. **Ahmed, M., and D. D. Focht.** Degradation of polychlorinated biphenyls by two species of *Achromobacter*. Can. J. Microbiol. **19**:47–52.
2. **Anonymous.** 1979. The biodegradability and bioaccumulation of new and existing chemical substances. 5.21, C65/79/JAP. Chemical Products Safety Division, Basic Industries Bureau, Ministry of International Trade and Industry, Japan.
3. **Astill, B. D., C. J. Terhaar, R. L. Raleigh, and A. N. M. Nasr.** 1977. A tier testing scheme. Health, Safety and Human Factors Laboratory, Eastman Kodak Co., Rochester, N.Y.
4. **Baxter, R. A., P. E. Gilbert, R. A. Lidgett, J. H. Mainprize, and H. A. Vodden.** 1975. The degradation of polychlorinated biphenyls by microorganisms. Sci. Total Environ. **4**:53–61.
5. **Bengtsson, B.-E., O. Svanberg, and E. Linden.** 1979. Chlorinated paraffins in bleaks (*Alburnus alburnus* L.). Ambio **8**:121–122.
6. **Bourquin, A. W., and S. Cassidy.** 1975. Effect of polychlorinated biphenyl formulations on the growth of estuarine bacteria. Appl. Microbiol. **29**:125–127.
7. **Furukawa, K., and F. Matsumura.** 1976. Microbial metabolism of polychlorinated biphenyls. Studies on the relative degradability of polychlorinated components by *Alkaligenes* sp. J. Agric. Food Chem. **24**:251–256.
8. **Furukawa, K., K. Tonomura, and A. Kamibiyashi.** 1978. Effect of chlorine substitution on the biodegradability of polychlorinated biphenyls. Appl. Environ. Microbiol. **35**:223–227.
9. **Howard, P. H., J. Santodonato, and J. Saxena.** 1975. Investigation of selected potential environmental contaminants: chlorinated paraffins. EPA-560/2-75-007. Office of Toxic Substances, U.S. Environmental Protection Agency, Washington, D.C.
10. **Jacobs, M. B.** 1949. The analytical chemistry of industrial poisons, hazards and solvents. Interscience Publishers, Inc., New York.
11. **Keil, J. E., S. J. Sandifer, C. D. Graber, and L. E. Priester.** 1972. DDT and polychlorinated biphenyl Aroclor 1242. Effects of uptake on *E. coli* growth. Water Res. **6**:837–841.
12. **Langley, E.** 1978. Health and safety aspects of the proposed notification scheme for new chemicals. Chem. Ind. (London), p. 504–507.
13. **Meylan, W. M., and P. H. Howard.** 1977. Chemical input/output analysis of selected chemical substances to assess sources of environment contamination. Task II: biphenyl and diphenyl oxide. EPA 560/6-77-003. U.S. Environmental Protection Agency, Washington, D.C.
14. **Sasaki, S.** 1978. The scientific aspects of the chemical substances control law in Japan, p. 283–298. *In* O. Hutzinger, L. H. Van Lelyveld, and B. C. J. Zoeteman (ed.), Aquatic pollutants transformation and biological effects. Pergamon Press, Inc., Elmsford, N.Y.
15. **Tucker, E. S., V. W. Saeger, and S. O. Hicks.** 1975. Activated sludge primary biodegradation of polychlorinated biphenyls. Bull. Environ. Contam. Toxicol. **14**:705–713.
16. **Wong, P. T. S., and K. L. E. Kaiser.** 1975. Bacterial degradation of polychlorinated biphenyls. II. Rate studies. Bull. Environ. Contam. Toxicol. **13**:249–256.

Use of a Microcosm Approach to Assess Pesticide Biodegradability

FUMIO MATSUMURA

Pesticide Research Center, Michigan State University, East Lansing, Michigan 48824

Biodegradation of pesticidal chemicals is not a matter that can be simply measured and expressed, particularly when the term is applied to all the chemical changes which occur in the environment. Changes in the molecular structure of pollutants in the environment are often brought about by many interacting "weathering" forces, including physical, chemical, and biological factors. The use of properly designed microcosm experiments to examine and predict the significance of these interactions is discussed, and specific research plans, ranging from tests with pure cultures of microorganisms to model ecosystem experiments, are provided.

There is no question that the overall persistence of pesticidal chemicals in the environment is largely determined by their biodegradability. Since persistence is often the cause of environmental problems in pollution cases, no one argues about the importance of developing proper test methods to screen the biodegradability of chemicals. Preferably, such screening efforts should be made on all new chemicals before they are introduced into the environment. The problem is that biodegradation is not a matter that can be simply measured and expressed, particularly when the term is applied to all the chemical changes which occur in the environment. Changes in the molecular structure of pollutants in the environment are often brought about by many interacting "weathering" forces, including physical forces such as sunlight irradiation, rain, etc. If one counts other secondary influences such as location and the state of the chemical, i.e., in bound, colloidal, or solubilized forms, all of which eventually affect the degradability of the chemical in nature, the degree of complexity of the problem can enormously increase. Certainly, it is true that the actual rate of degradation of any chemical varies from one environment to another. Even within the same locality, the speed of degradation may vary appreciably by the season, the distribution pattern of the chemical, and differences in conditions of microenvironments. For instance, the half-life of toxaphene has been quoted as 1 or 2 days on range land (16) in the summer to over 20 years in a model ecosystem study (14). Thus, in assessing the biodegradability of chemicals one must first know that there is no perfect environment where one can obtain reproducible quantitative data. Even under a strictly controlled experimental condition such as experimental farms, persistence data may vary from one lot to another and from year to year. One of the first things we learn in this field is that "biodegradability" (= biodegradation potential) is a term applied only to relative relationships among compounds with respect to their liability in nature.

MICROBIAL DEGRADATION OF PESTICIDES

When pesticides are introduced into the environment, the bulk of them come in contact with soil, including aquatic sediment, whether or not they are purposely aimed at it. Indeed, soil is regarded as the largest reservoir of these chemicals, from which they may gradually dissipate, be transported, and accumulate in other systems. It is generally agreed that two major degradation forces in the environment are sunlight and biological systems. Among biological systems, microorganisms are known to play by far the most important role. The reason for this is multifold: first, the locations (i.e., soil and aquatic sediment) where the bulk of these chemicals end up in the environment; second, the surface area they represent; third, the genetic variability of microorganisms combined with their high metabolic activities; and fourth, their ability to metabolize even the most recalcitrant chemicals under many environmental conditions. This is not to say that plants and animals do not also play important roles in pesticide degradation; certainly, in given situations such as food chain accumulation, their roles cannot be overlooked. Also, plants, because of their large biomass, play a significant role in altering the fates of pesticides. However, one must consider the fact that many of the plants and animals may be regarded as the "victims" of these environmental pollutants (e.g., nontarget species in the agricultural environment, wildlife, etc.), and as such many scientists are hoping that the chemicals they are concerned with will degrade in the environment before they reach these organisms.

As to the degree of contribution from sunlight and microorganisms to the overall degradation of pesticidal chemicals, one can naturally assume that the location of the chemicals with regard to the availability of sunlight is the most important factor. Thus, chemicals evaporated or otherwise floating in the air (e.g., dust-carried transport) are the ones most likely to be attacked by sunlight. The most active region of sunlight, ultraviolet ray, is known to penetrate into water to various depths, depending upon the dissolved and suspended material in water (e.g., a few decimeters to a few centimeters). Pesticides located on the surface of soil and plants are also subjected to photochemical attacks.

As to the possibility of simulating sunlight, today's technology is such that light sources which resemble the ultraviolet spectrum of sunlight are readily available (17). The only remaining problem is that in the environment various substances affect both the nature of the reaction and the availability of the substrate to light. In the former case, they could not only affect the rate of reaction (e.g., photosensitizing, inhibition) but also change the nature of reaction products.

MICROCOSMS APPROACH

Microcosms, by definition, are experiments designed to simulate or study the events that actually take place in the environment. Much has been said about microcosms, and several symposia have been held to discuss the pros and cons of the microcosm approach.

The problems in utilizing microcosm approaches have been (i) the existence of too many different designs, each suiting some purpose; (ii) the inclusion of scientists from many, often unrelated, fields of science, such as system scientists versus chemists; (iii) the lack of the realization that no microcosm can ever produce the same overall effects as the environment; (iv) the lack of reproducible experimental design; and (v) disproportional biomass in microcosms as opposed to the natural environments. Above all, item (iii) shown above appears to be the most serious problem. In the past, certainly too much was expected of microcosms. Some people have expected absolutely quantitative microcosm data which are identical to data obtained from nature. Other people have tried to put everything together in the hope of simulating all the reactions that go on in the environment, only to find that such complex systems are too difficult to reproduce, they are too cumbersome to allow control of each environmental factor, they are too costly, and, most serious of all, they yield data which are very difficult to interpret.

On the other hand, one must look at the problem with a positive attitude. After all, there is no panacea in any field, and yet progress has been made in many fields by agreeing on a uniform initial approach. Toxicologists, for instance, have been relying heavily upon the use of rats as the primary screening organism, though rats do not give all pertinent toxicity data on human and other important biological systems. Molecular biologists use *Escherichia coli* for their studies on basic cellular function despite the fact that it is not even a eucaryotic organism. These approaches have served as the primary study protocol for scientists in each field. Other secondary and tertiary study materials, such as primates and human cell and organ cultures for the toxicologist and *Drosophila* for the molecular biologist, are used to supplement the deficiencies caused by the choice of the primary studies.

At the same time, if any research team is to make positive contributions, before devising a plan of approach, they have to clearly understand microcosm studies, the philosophy behind them, the purpose, and the strong points and the weaknesses of microcosmology. *Philosophy:* Any microcosm system is designed to study merely a small facet of environmental effects. Often, only one aspect of environmental events can be studied or simulated in one set of experimental design. While it is possible to design the experiment to ask questions on specific issues (e.g., photodecomposition, etc.), it must be clearly understood that no microcosms would give data quantitatively equal to the events taking place in nature. All values generated by model ecosystems-microcosms are therefore valid only in the relative sense. *Purpose:* Microcosm studies may be used for either screening or prediction purposes so long as their limitations as outlined above are understood. However, microcosms do not eliminate the necessity of field testing or environmental studies. Microcosm studies may be a prominent tool for screening many premarket chemicals, inasmuch as they would give manufacturers the sense of relative stability and bioaccumulation potentials for compounds, and thereby aid in making the decision of whether or not to further pursue a course.

MICROCOSMS: DEFINITION AND EXPLANATION

Many people, including scientists themselves, have had trouble defining microcosms. Microcosms mean one thing to one section of science and another thing to others. There are basically four types of microcosm approaches:

1. *"Bench mark"* approaches. Physiochemical parameters such as volatility, water solubility, partition coefficients (e.g., water-octanol partitioning), etc., are determined and utilized to predict the probability of environmental behavior of compounds.

2. *Single-reaction approaches.* Only one set of interactions between the

compound and one of the environmental forces, either biological or physical, is studied. The use of microbial pure cultures, photo-illumination to simulate sunlight, fish accumulation studies in the absence of food organisms, etc., are included in this category. The advantages of such systems are (i) reproducibility, (ii) ease by which various environmental factors can be controlled and varied, and (iii) usual indication of the presence of the cause-effect relationship.

3. *Multicomponent systems.* More than one factor is placed in one system. Multicomponent systems range from a small test tube containing nonsterile water to complex self-sustaining model ecosystems which may house various organisms at different trophic levels in constituting an ecosystem-food web. Often, only multicomponent systems are referred to as microcosms. The advantages of this type of microcosm are (i) they are one step closer to the environment, (ii) often the data on the ecological interactions can be generated, (iii) multistep reactions can be studied, etc. On the other hand, the disadvantages of a multicomponent system have already been described in this paper.

4. *Systems analysis approach.* This route of approach to environmental studies has been made by mathematically oriented scientists such as systems analysts, biostatisticians, etc. In many cases, these scientists use the data generated by model ecosystems, small field studies, or laboratory experimentation to predict or determine the rate at which compounds are metabolized and/or transported through the biosphere.

CONSIDERATIONS ON APPLICATION OF MICROCOSM TECHNOLOGY TO MICROBIAL STUDIES

Relative nature of microcosm data. It is most imperative to understand the fact that the microcosm approach can only give values which are relative to other materials tested in the same system. That is, microcosms may give the data that one compound, say DDT (dichlorodiphenyltrichloroethane), degrades slower than another similar compound, say methoxychlor, in the same experimental setup. The system by itself does not necessarily give the actual value as to how many months or years the compound lasts in nature. Microcosms could, however, tell whether compound A is more degradable than B under given conditions.

Reproducibility of the data. To ensure the validity of microcosm experiments, it is imperative that the given system produce reproducible data within a set of experiments. Otherwise, comparison among compounds as proposed above would become meaningless. It must be pointed out here that such a requirement tends to favor small and simple systems over more complex ones. The latter systems certainly have a place in areas other than the comparative studies as outlined above. Proper controls, such as a sterilized system for microbial studies, are also to be used to ensure that no artifacts are involved.

Consideration of biomass. Any model ecosystem which contains disproportional amounts of biomass could give unrealistic data. For instance, if a large animal is placed in a small model system, the animal could consume most of the test compound in a relatively short time span and thereby minimize the differences among test compounds. In this context, I believe that microbial studies (including some algal groups) are most suited for model ecosystems because of their small size in relation to that of the model system.

"Bench mark" approaches are a prerequisite for model ecosystem studies. There is no question that physical properties do affect the fate of chemicals in the environment. The choice of type and design of model ecosystems could be guided by the physiochemical parameters obtained by bench mark approaches. For instance, if the test compound is known to have a high volatility, studies on air-mediated transport and photochemical degradation would be the logical choice. If the compound is shown to be lipophilic, tests on its bioaccumulation potential are going to be necessary. In the case of microbial studies, potential of cellular pickup, mode of contact (e.g., water mediated), molecular size and composition of chemicals, etc., are important to consider.

Adoption of "tier" systems. Since each microcosm approach can only look at limited aspects of environmental events, it is more desirable from the viewpoint of a screening to test several simplified microcosm approaches rather than to rely upon one complex and elaborate model in the hope of getting many answers all at once. These can be open or closed systems and must include physiochemical controls.

Typically, a screening process could begin with bench mark approaches and then proceed to simplest test systems, such as degradation studies in pure cultures of microorganisms appropriate to the recipient system, or to the use of more complex systems, such as degradation studies using soil plugs, mixed cultures of microbes, lake sediments, etc. In due course, radiolabeled compounds must be used in the more complex model systems. Efforts must be made to account for all radioactivities, including CO_2 and other volatilized portions. At the last step, appropriate complex model ecosystems may be tested. This step is not meant to obtain relative indices among various test compounds, but to study ecological significance.

Field or semifield testing. In the final analysis, field testing is an important step in ascertaining the environmental safety of the chemical. The compounds which have been advanced to this stage must show great economic promise. In the case of metabolic studies, this is the step whereby the presence of the metabolites identified by the model systems is confirmed and their quantities and environmental significance are determined. Indeed, many environmental conversion products have been identified in this manner. At the same time, it must be pointed out that in several cases some surprises were also there: certain reactions which looked prominent in microcosm studies may not have been found in the actual field test. Thus, the value of small-scale field testings cannot be overemphasized. At the same time, this is the stage at which the benefits of data organization from mathematical model developments may be realized.

RESEARCH PLANS

Tests by Using Pure Culture of Microorganisms

A typical research program on microbial degradation of pesticidal chemicals should begin with pure cultures of microorganisms that are readily available. In our laboratory we have used specific cultures of *Pseudomonas putida* (ATCC

17484) and *Trichoderma viride* (ATCC 13631) for such a purpose. The choice of these organisms is not entirely empirical. They are commonly occurring in soil, and each represents a large group of organisms which are well studied as to their basic physiology.

Pseudomonas species in particular are very common bacteria in soil. They possess tremendous versatility in metabolizing various organic compounds. *P. putida* is, moreover, known to be capable of utilizing aromatic and cyclic compounds via ring-opening actions. Pseudomonads are basically aerobic organisms. However, they can also stand mild levels of anaerobic conditions where basic reductive degradation such as dechlorination of DDT (12) and toxaphene (3) may be carried out. As to the choice of medium, we have found the No. 2 medium of Fred and Waksman (4), consisting mainly of mannitol and yeast extract, to be a good one in general screening.

T. viride is a common soil fungus. It is known to possess a variety of metabolic capabilities (11). One important aspect of this species is that it is known to produce hydrolases (9) which degrade many types of chemicals via hydrolysis. Production and release of such hydrolases are considered to be the mechanism by which these organisms first break up large molecules, probably outside of the cell (i.e., exoenzymes), so that they can be later taken through their cell membrane. Exoenzymes have been considered to be the cause of degradation of many pesticides, especially organophosphates and carbamates in soil; thus, the use of *T. viride* as a test organism may help scientists in relating biodegradability to the action of general hydrolytic enzymes.

In general, degradation studies must be conducted in two sets of conditions, one oxidative (shake culture) and the other reductive (stagnant culture). This is because of the possibility that the test compound might behave in very different ways in these two sets of conditions. The substrate is added toward the end of the growth period (i.e., log phase) when the total number of cells stabilizes. One must be careful not to overload the system, however, by adding a large quantity of the substrate. That is, pesticide residues generally occur at low levels, typically in parts per million and parts per billion ranges. Degradation as expressed in half-life is a function of concentration. Thus, conditions must be selected so that the levels of the substrate reflect the conditions in nature or the degradation capability of the organisms, or both. At the same time, any bacteriostatic or fungicidal effects of the substrate must be noted. In all cases standard degradation activities of the cultures under the given experimental conditions must be measured against known substrates such as DDT, malathion, 2,4-D (2,4-dichlorophenoxyacetic acid), and parathion. These are the chemicals whose biodegradabilities in the environment are generally known as a result of many years of research work. So far as possible, a series of related chemicals should be chosen so that general structure-degradability relationships will be known. It must be mentioned here that the correlations between laboratory data and field data have been found to be most positive in the case of tests on related compounds such as methoxychlor versus DDT, 2,4-D versus 2,4,5-T (2,4,5-trichlorophenoxyacetic acid), etc.

Tests on selected microbial cultures. In some cases, the above "pure culture" systems may not degrade the test chemical. This could mean either that the test compound has a low biodegradability or that these two species of microorganisms under the test conditions happen to be incapable of metabolizing this substrate. Two types of approaches may be made in such cases. First, more microbial isolates from a variety of sources may be tested. Sediment samples from lakes and rivers or soil samples from farms and gardens where the pesticidal chemicals are likely to be used or end up would be generally good enough. In certain cases where difficulties of biodegradation are suspected, a search of sources where the chemical has been known to be present for some time may yield degrading microorganisms. One important fact to remember is that the choice of culture medium affects the outcome of such tests. Thus, the use of specialized media must be avoided, since they would pick up only certain types of organisms from the source. Again, general media such as No. 2 medium of Fred and Waksman and potato-dextrose agar are recommended for this purpose. By this approach we have found microorganisms metabolizing dieldrin (10), endrin (13), and chlordane (R. W. Beeman and F. Matsumura, unpublished data), for instance. There is a general tendency to correlate the ease with which one finds an active organism and the degree of biodegradability of the compound. To cite the same example, the ratio of success to failure in finding an active metabolizer was roughly 1 to 400 for dieldrin, 1 to 100 for endrin, and 1 to 75 for chlordane. Certainly, this way of estimating degradability is not precise, nor is it universally applicable, but if it is used as one of the ways to assess biodegradation potential and is combined with other evidence, it often gives an early indication of degradation tendencies.

Another method utilized to find microbial strains which degrade stable organic chemicals is a cometabolic approach (2, 8) whereby mixed microbial sources are given an opportunity to grow on pesticides or their degradable analogs as sole carbon sources. The technique is also called the "enrichment culture" technique. One good example is the approach used by Ahmed and Focht (1) on polychlorinated biphenyls. These workers cultured microorganisms which grow on biphenyl (non-chlorinated) or monochlorobiphenyl as sole carbon sources. These selected microorganism strains were then given the chance to metabolize polychlorinated biphenyls. This way, even very stable types of chemicals were found to be degradable by microorganisms.

Degradation of Pesticidal Chemicals by Model Ecosystems

The above studies using pure isolates of microorganisms are helpful in establishing the fact of microbial metabolism, and give one an insight into the metabolic processes and the nature of metabolic products. They do not assure that such active metabolizers indeed remain active in soil and sediment to carry out the task. Model ecosystems involving microorganisms are a valuable tool in this regard. Also, such an approach gives one an opportunity to examine situations close to the site where the pesticides or related chemicals are likely to end up.

Typical examples of such attempts can be found in the literature. In the experiment used by Lee and Ryan (7), for instance, a small quantity of sediment from a local estuary is added to water in an appropriate-sized flask, and after the pesticidal substrate is added, the system is simply shaken to promote aerobic metabolism. Another way to provide a somewhat aerobic environment is to use air bubbling to sediment-water systems (15). Some nutrients may be added to the system, as they promote microbial activity. Cometabolic activities are promoted by this approach, but not true metabolism where pesticidal substrates are used as carbon sources by microorganisms. Utilization of activated sludge is also based upon the same line of logic and should be of particular value to scientists who study industrial and municipal waste problems.

Along with the above aerobic study, a simultaneous anaerobic study should be conducted on the same test compounds. Static or stagnant systems involving sediments and water provide an anaerobic microbial environment. Stagnant laboratory aquaria or test tubes with deep enough sediment and water, for instance, easily establish an anaerobic environment in the neighborhood of -200 to -300 mV in a few days (6). Microbial populations tend to stabilize after a few weeks in the laboratory and thereby provide stable sources, if aquatic sediments are once collected in large enough quantities and kept in a controlled environment. In some cases, addition of extra nutrients provides high microbial activities and anaerobic conditions.

Specific research plan. In order to accommodate all types of metabolic forces that are likely to occur in pesticides and related chemicals, the following six-step incubation procedures are adopted.

Step 1. Establish large reservoirs of sediment, either from sludge or a lake or river, and water in the laboratory. Keep them at a constant temperature (usually 24°C, in some cases 4°C, depending upon the purpose) and light. After 1 week of storage, the sediment and water are ready for the experiment.

Step 2. To a 250-ml Erlenmeyer flask, 100 ml of river or lake water and 3 g of sediment are added. The test chemical is added as a dry coating (e.g., use acetone as a vehicle and evaporate it to leave the chemical on the sand surface) on 1 g of clean sand. Stir the system to disperse the chemical evenly. Thereafter, let the system settle down and close the top with a cotton plug.

Step 3. First store the flask for 48 h without shaking (i.e., stagnant condition) at 24°C; then transfer the flask to a rotary shaker and shake for additional periods. The duration of this shaking is dependent upon the nature of the compound. For a new system it is recommended that samples be taken from time to time to monitor the progress of degradation. In any case, terminate the experiment after 28 days.

Step 4. Extract the entire system by using appropriate solvents. It is preferable that the extraction efficiency of the parent compound be checked prior to the experiment. For non-water-soluble materials, the use of a universal solvent mixture such as chloroform-methanol (e.g., 5:1) may be tried.

Step 5. As the first step of identification, the level of the parent compound remaining in the system should be appropriate for a quick screening purpose.
Other conditions. In all cases, well-selected *standard chemicals* (= "bench mark" chemicals) must be used to indicate how the system is operating. For general screening, chemicals such as DDT, parathion, 2,4-D, benzene hexachloride, etc., whose general environmental fates are well known should be used. Also, at least two types of control

(= blanks) must be used along with the above degradation test. They are a dark control (the same experiment in the dark) and a sterile control. Autoclaving is an acceptable method of sterilization at this level, though for more elaborate studies chemical (e.g., propylene oxide) and radiation sterilization methods may have to be considered. The choice of concentration (expressed as parts per million, parts per billion, parts per thousand, etc., against the total volume of sediment plus water) must reflect the probable levels of environmental contamination by the test chemical. For instance, 100 ppm or a higher level would be needed to investigate the effect of point source contamination, whereas 1 ppm or even 1 ppb levels are appropriate for most non-point source pesticide contamination. Initial testings could be done at only one concentration, but tests of at least one additional level of concentration would give valuable information on the behavior of the chemical in the environment.

Step 6. If the above experiments fail to give a positive indication that the compound in question is degradable, an enrichment technique should be tried. This can be done by adding the test compound to a larger system (e.g., 10× scale) at the same substrate concentration as the standard test prior to the actual testing. At the same time, add mineral supplements to the water to promote microbial activities. However, no extra carbon sources should be added to the system. Start the regular incubation process (i.e., shaking at 24°C under constant laboratory lighting). The period of preincubation should be usually on the order of 10 days, but there may be examples where incubation of up to 1 month would be required to acclimatize the system. Proper care must be taken to correct for the level of the residual substrate at the beginning of the regular testing.

Whatever the approach is, it is most imperative that the reproducibility of the degradation activity be checked by using standard substrates. Also, throughout model ecosystem studies, it is important to establish that the observed degradation activities are due to microbial action by using appropriate sterile blank tests. The use of autoclaved systems has been criticized in the past since such treatments tend to destroy both exoenzymes (and soil enzymes) and delicate soil structure which could affect soil degradation of pesticides and microbial activities. In the case of intact soil plugs in particular, both gamma radiation and chemical treatments (5) have been used for this purpose. It must be emphasized here again that all precautions mentioned for pure-isolate studies (e.g., dose, use analogs, etc.) are necessary to carry out such model ecosystem studies involving microorganisms.

The initial model ecosystem for photochemical degradation should simply involve an appropriate phototube to simulate sunlight and pure medium such as n-hexane. This solvent is known to act as a proton donor, but since there are enough such donors in most environments, its choice may be defended. Water is another medium that should be used for photochemical reaction. It must be mentioned here that various types of dissolved matter and ions affect photochemical reactions, and nucleophilic substances such as OH- (e.g., at high pH's) and CN$^-$ in particular promote photonucleophilic reactions. If one has access to stable sources of water such as lakes and rivers, model ecosystems for photodegradation based upon the local source of water would be of great use in predicting the photochemical fate of pesticides in those environments.

LITERATURE CITED

1. **Ahmed, M., and D. D. Focht.** 1973. Degradation of polychlorinated biphenyls by two species of *Achromobacter.* Can. J. Microbiol. **19:**47–52.
2. **Alexander, M.** 1979. Role of co-metabolism, p. 67–75. *In* Microbial degradation of pollutants in marine environments. EPA-600/9-79-012. Environmental Protection Agency, Gulf Breeze, Fla.
3. **Clark, J. M., and F. Matsumura.** 1979. Metabolism of Toxaphene by aquatic sediment and camphor-degrading pseudomonad. Arch. Environ. Contam. Toxicol. **8:**285–298.
4. **Fred, E. B., and S. A. Waksman.** 1938. Laboratory manual of general microbiology. McGraw-Hill Book Co., New York.
5. **Getzin, L. W., and I. Rosefield.** 1968. Organophosphorus insecticide degradation by heat-labile substances in soil. J. Agric. Food Chem. **16:**598–601.
6. **Isensee, A. R., P. C. Kearney, and G. E. Jones.** 1979. Modeling aquatic ecosystems for metabolic studies, p. 195–216. *In* M. A. Q. Khan, J. J. Lech, and J. J. Menn (ed.), Pesticide and xenobiotic metabolism in aquatic organisms. Am. Chem. Soc. Symp. Ser. 99. American Chemical Society, Washington, D.C.
7. **Lee, R. F., and C. Ryan.** 1979. Microbial degradation or organochlorine compounds in estuarine waters and sediments, p. 443–450. *In* Microbial degradation of pollutants in marine environments. EPA-600/9-79-012. Environmental Protection Agency, Gulf Breeze, Fla.
8. **Matsumura, F., and H. J. Benezet.** 1978. Microbial degradation of insecticides, p. 623–667. *In* I. R. Hill and S. J. L. Wright (ed.), Pesticide microbiology. Academic Press, London.
9. **Matsumura, F., and G. M. Boush.** 1966. Malathion degradation by *Trichoderma viride* and a *Pseudomonas* species. Science **153:**1278–1280.
10. **Matsumura, F., and G. M. Boush.** 1967. Dieldrin: degradation by soil microorganisms. Science **156:**959–961.
11. **Matsumura, F., and G. M. Boush.** 1968. Degradation of insecticides by a soil fungus, *Trichoderma viride.* J. Econ. Entomol. **61:**610–612.
12. **Matsumura, F., and E. G. Esaac.** 1979. Degradation of pesticides by algae and aquatic microorganisms, p. 371–387. *In* M. A. Q. Khan, J. J. Lech, and J. J. Menn (ed.), Pesticide and xenobiotic metabolism in aquatic organisms. Am. Chem. Soc. Symp. Ser. 99. American Chemical Society, Washington, D.C.
13. **Matsumura, F., V. G. Kanvilkar, K. C. Patil, and G. M. Boush.** 1971. Metabolism of endrin by certain soil microorganisms. J. Agric. Food Chem. **19:**27–31.
14. **Nash, R. G., and E. A. Woolson.** 1967. Persistence of chlorinated hydrocarbon insecticides in soil. Science **157:**924–927.
15. **Pritchard, P. H., A. W. Bourquin, H. L. Frederickson, and T. Maziarz.** 1979. System design factors affecting environmental fate studies in microcosms, p. 251–272. *In* Microbial degradation of pollutants in marine environments. EPA-600/9-79-012. Environmental Protection Agency, Gulf Breeze, Fla.
16. **U.S. Department of Agriculture.** 1959. Residues in fatty acid, brain, and milk of cattle from insecticides supplied for grasshopper control on rangeland. J. Entomol. Soc. Am. **52:**1206–1210.
17. **Zabik, M. J., R. A. Leavitt, and G. C. C. Su.** 1976. Photochemistry of bioactive compounds: a review of pesticide photochemistry. Annu. Rev. Entomol. **21:**61–79.

Synopsis of Discussion Session: Case Studies

H. E. JOHNSON, V. ZITKO, AND F. MATSUMURA

The papers presented on the case studies of nitrilotriacetates, polychlorinated biphenyls (PCBs), and pesticides show that past experience with these compounds has not provided useful guidelines for hazard evaluation of new chemicals. The studies on these chemicals did not follow an established logical order, but rather were dictated by a succession of events. On the other hand, the amount of information accumulated on these chemicals provides a valuable data base which is useful for comparison with other chemicals for which only limited information is available.

As with most environmental data currently in existence for the majority of chemicals, the information was developed in response to individually dictated or specific needs. To date, very few chemicals have been subjected to a systematized or sequential hazard evaluation program designed to develop integrated data on environmental fate and biological effects. As evidenced by the case studies of these three classes of chemicals, this lack of overall guidance for needed hazard evaluation data has resulted in the generation of much unrelated data, certainly of interest to site- or system-specific applications but of marginal utility with respect to overall environmental hazard evaluation needs.

Environmental studies on the behavior of pesticides, PCBs, and related halogenated aromatics are probably more extensive than those with any other group of chemicals. Enormous amounts of data on residue accumulation and translocation have been generated which should be valuable for assessing the large-scale movement and translocation of these types of chemicals.

Similarly, there has been extensive investigation of pesticides and related compounds in the area of microbial metabolism, physiochemical bases for persistence, and the behavior of these chemicals in soil and water. Thus, in studying the environmental fate of new chemicals, it is important to utilize these data which are already available.

The value of these data may be enhanced by obtaining some additional data according to the procedures outlined in this and preceding hazard evaluation workshops. Additionally, some of the existing laboratory data may be verified in laboratory microcosm studies. Such validation studies would be useful to assess and correct errors in previous laboratory experiments.

APPLICATIONS OF BENCH MARK CHEMICALS IN HAZARD EVALUATION

The comparison of biological effects and environmental fate data for a new chemical with previously existing data for well-studied chemicals is termed the "bench mark approach." Using existing data as a bench mark with which to compare newly generated data for new or poorly studied classes of compounds offers an early qualitative assessment that can direct further studies. Structure-activity correlations have been used in this manner for predicting toxicity and bioconcentration potential.

The use of bench mark chemicals and the relative ranking of chemicals for the purposes of hazard evaluation did raise some objections from the workshop participants. Most of the objections were aimed against the use of DDT (dichlorodiphenyltrichloroethane) as a bench mark chemical, because of its unique chemical structure. In addition, strong doubts were expressed over the applicability of any relative ranking system to the estimation and prediction of environmental exposure concentration. After a discussion, a consensus was reached that bench mark chemicals are useful as internal standards and that relative ranking of chemicals has a place in hazard evaluation. Thus, for example, an organophosphate pesticide such as parathion could serve as a bench mark for the evaluation of other organophosphate pesticides, carbaryl could serve for carbamates, PCBs for polycyclic aromatic hydrocarbons in respect to chemical properties of these compounds, nonyl phenol for longer-chain (more than one carbon) alkyl phenols, etc. The choice of relatively well-studied chemicals as bench marks takes advantage of the wealth of data available for these compounds, both from laboratory studies and from field experience, and may allow the extrapolation from laboratory measurements to environmental situations for other chemicals. The extrapolation may initially be only qualitative, but it is not unreasonable to expect that a data base may be developed that would allow quantitative predictions in a manner analogous to quantitative structure-toxicity or bioaccumulation relationships for rates of biotransformation and transport, and other parameters as well.

Comparative studies emphasize similarities and differences between chemicals. Such direct comparisons are likely to provide a deeper insight into the mechanisms underlying the various processes determining environmental behavior of chemicals. In addition, chemicals that, on the basis of laboratory tests, appear to have more significant environmental exposure concentrations than the appropriate bench mark chemical would likely receive immediate attention. As an example, a qualitative comparison of high-molecular-weight chlorinated paraffins to PCBs is outlined in Table 1.

The comparison indicates that the degree of environmental contamination by chlorinated paraffins is likely to be lower than that by PCBs (probably no emissions of chlorinated paraffins from incinerators, no bioaccumulation of C-24 chlorinated paraffins), but a low-level contamination of the physical environment by both types of chlorinated paraffins, and of biota by C-12 chlorinated paraffins, is likely. This contamination, combined with the lack of data on chronic toxicity and biodegradation, underlines the importance of further studies in these areas for hazard assessment of chlorinated paraffins and the need for better analytical methods to support these studies.

PESTICIDE DATA AND THE USE OF MICROCOSMS

Without doubt, pesticides, PCBs, and related halogenated aromatics represent one of the most researched and documented group of chemicals from the viewpoint of environmental studies on the behavior of chemicals.

In the case of pesticides, their earlier studies were confined to behavior in agroecosystems, particularly with respect to residue characteristics in food crops, since the Food and Drug Administration mandates require residue data for registering pesticides. More recently, as a result of concerns over envi-

TABLE 1. *Example of the bench mark approach comparing the relatively well-studied PCBs with C-12 and C-24 alkyl chain-length chlorinated paraffins*

Factor	PCBs	Chlorinated paraffins
Production	Same order of magnitude	Same order of magnitude
Uses	Closed, open	Mostly open (plastics, paints, lube oils)
Heat stability	Very stable	Decomposition at 300°C
Chemical stability	High	Less stable than PCBs
Solubility in water	Low	Low
Lipophilicity	High	High
Acute toxicity	Low	C-12 alkyl stock high to fish; C-24 alkyl stock low
Chronic toxicity	High	No data
Biodegradation	High to nonbiodegradable; decreases with increasing chlorine content	Unknown; probably low
Bioconcentration	High	C-12 alkyl stock intermediate; C-24 alkyl stock very low
Analytical technique	Routine	Not well developed

ronmental and health effects, enormous amounts of data dealing with residue characteristics and dynamics of translocation of pesticidal chemicals have been generated. Although some of the data are of limited value from the viewpoint of biotransformation studies, there is much useful information that can be obtained from these fragmented studies. Particularly important are the large-scale and long-term monitoring data on the changes of residue levels and characteristics of such well-studied compounds as DDT, dieldrin, PCBs, etc., before and after the banning of their use. These data are indispensable in assessing the large-scale movement and transformation of this type of chemicals, for no laboratory experiments can yield such information.

Narrowing the discussion to the cases related to microbial transformation, one is initially surprised by the wealth of information in the area of microbial metabolism, enzymatic bases of metabolism, physicochemical characteristics of chemicals that affect persistence, behavior of these chemicals in soil and soil microbial ecology, and residue data in many different ecosystems. Thus, in studying the environmental fate of new chemicals, it is imperative that we utilize these potentially useful data already available. The question then becomes one of identifying the most promising methods and approaches for the generation of needed data.

In this regard, "microcosm approaches" appear to be one of the methods which show a good promise, and, therefore, their merits and limitations are discussed below.

Microcosm Approaches

Though there are many different opinions as to what microcosm approaches really are, most people would agree that they represent efforts to study complex environmental events by simplification through reduction of the number of interacting factors and/or miniaturization of a specific component so that definite cause-result relationships can be studied in the laboratory. In the opinion of many scientists, microcosm studies should also include the efforts of validation (i.e., check the data generated by the above laboratory studies in actual environments, such as comparing the laboratory data against residue data or the data generated by field experiments). In turn, the result of such comparisons can be used to correct the errors generated by the laboratory experiment (i.e., fine tuning).

It must be stressed here that applied microcosm studies must include the stage not only of simplification/miniaturization but also of validation processes to be included under actual use conditions or in the field. Whatever the method of approach, the study must represent an effort to understand the environmental events and to duplicate at least some aspects of the environment as realistically as possible. Within this cycle (i.e., from studies in environmental events to laboratory phenomena and back to environmental events) is a specific phase where some degree of prediction possibility may be realized. This happens only when there are already adequate amounts of environmental data available, and when enough reproducible data have been generated in the laboratory and at least one control factor has been clearly established (e.g., microbial role, certain chemical properties affecting biotransformation, etc.). Preferably, the system must also be tested against enough variables, such as a number of compounds or different biological materials and conditions (i.e., base-line data), before it is used in a predictive model. Above all, it must be made clear once more that even the most established predictive model should be validated through reexamination of the generated result against the events which occur in nature.

Chapter 8. SUMMARY, CONCLUSIONS, AND FUTURE RESEARCH NEEDS

Chairman: J. L. HAMELINK

Committee Members: D. T. GIBSON, G. F. LEE, G. V. LOEWENGART, A. M. STERN, AND J. M. TIEDJE

HISTORICAL REVIEW

"Hazard Evauation of Chemicals in Aquatic Environments" has been the central theme for three successive summer workshops. The first workshop (1) focused on the effects of chemicals on life forms and the general fate of chemicals in aquatic environments. A fundamental principle of hazard evaluation which came out of that workshop was the basic strategy of comparing the expected exposure concentration (EEC) to a measured no-observed-effect concentration. It was recognized that the quality of the comparison, and hence the confidence of any decision, was expected to improve as measurements proceeded from an array of simple through complex tests. Consequently, the concept of the "tier" approach was refined to provide a rational basis for making decisions at each successive level of information acquisition.

The second workshop (2) focused on the programs needed to obtain the information required for making hazard evaluation decisions. The policies of several organizations, including industrial groups and governmental agencies from around the world, currently being employed or under active consideration, were reviewed. Primary emphasis in the evaluation process was given to how well each program utilized the participant's collective understanding of basic hazard evaluation principles that had been developed the previous year. This naturally led to discussions concerned with application of the principles to specific environmental issues, notably the establishment of Water Quality Criteria by the U.S. Environmental Protection Agency. It should be recognized that the development and implementation of Water Quality Criteria is a popular topic because new criteria have been recently proposed for the so-called "toxic pollutants" and because innovative methodology has been used by the Environmental Protection Agency to relate fragmented information on relatively few chemicals to the evaluation of a large variety of chemicals. Utilization of the criteria to formulate site-specific Water Quality Standards received additional discussion which helped identify certain deficiencies in knowledge that could provide a basis for future workshops.

Biotransformation was selected as the subject for this workshop since it is often an important variable in the environment, yet uniform methods for generating and interpreting data are lacking. Biotransformation may become particularly important in a hazard evaluation process when the relative hazard is not clearly defined or if the margin of safety found for a particular situation is narrow. In these cases, precise determination of all the processes that affect

140

the EEC becomes critical. If a moderate rate of biotransformation is found to be significant, then the potential for complex interactions exists, resulting in substantial variations in the EEC in different receiving water bodies. Consequently, biotransformation as it relates to those hazard evaluations where the limits between effects and exposure are poorly defined emerged as the focus of this workshop.

GENERAL OBJECTIVES

The general objectives of this workshop reflect man's recognized need for maintaining and, where possible, improving environmental quality. An array of environmental legislation has been enacted by many countries over the past few years expressly to prevent deterioration of the environment by chemical pollutants. Both industry and regulatory agencies have to make hard decisions in order to meet the mandates of this environmental legislation. For example, industry has to arrive at early decisions regarding the probability that a new chemical will meet environmental constraints, because developmental costs for new products are very high. Similarly, regulatory agencies must process large volumes of data in a cost-effective manner in order to make timely decisions regarding potential risk posed by chemical releases. Although the decisions made by industry and government may address different objectives, many of the steps involved in their formulation are the same. Both groups have to (i) identify parameters that will lead to reliable decisions, (ii) identify methods that accurately measure biotransformation parameters, (iii) define the limitations of the methods regarding data interpretation and extrapolation to the environment, and (iv) define how the data will be used to make critical decisions.

The conduct of a safety assessment and the implementation of new environmental regulations require a reasonable system for assessing the potential risk posed by released chemicals. Any analysis of risk involves the integration of (i) the concentration of chemicals producing adverse biological effects and (ii) the concentration of chemicals expected or predicted to exist in the various environmental media such as soil, water, sediment, and the atmosphere. The latter subject is the area upon which this workshop focused its attention.

The workshop consisted of individuals from academia, industry, and government. More importantly, the group was multidisciplinary in make-up, and each participant was a recognized expert in his or her particular discipline. The organizers recognized that, if consensus was to be achieved, resolution of the many problems contained in the workshop subject area would require a multidisciplinary approach. The diversity of technical backgrounds possessed by the participants often resulted in healthy differences of opinion and frequently provided fresh approaches to "old" problems as the formal presentations and discussions took place during the week.

It is hoped that both the diversity and consensus of opinions expressed by the biologists, chemists, and physical scientists in attendance will carry through into this summary document so that the vitality of the workshop experience can be fully shared with the reader.

SPECIFIC OBJECTIVES

The workshop was responsible for covering varied aspects of biological transformation. These included the critical evaluation of methods for evaluating biotransformation, the utility of biotransformation data in the estimation of environmental concentration, and the problems associated with extrapolating laboratory data to natural environments. To accomplish this, the workshop was divided into six subgroups which addressed the following related and, in some cases, overlapping subject areas:

1. Exposure concentration
2. Evaluation of methods
3. The materials balance approach to environmental concentration
4. Environmental extrapolation of data
5. Quantitative expression of biotransformation
6. Case studies

The following section discusses the extent to which the above objectives were met by the workshop.

GENERAL CONCLUSIONS

"Biotransformation test methodology is now available, and the data derived from current laboratory practices can be extrapolated to various aquatic environments with sufficient reliability to be useful in hazard evaluation programs."

Biotransformation tests are used to assess the extent to which natural microflora can alter a chemical or mixture of chemicals. In order to contribute to the estimation of EEC, laboratory biotransformation methodology must be applicable or related to similar processes in natural environments. The workshop focused on numerous issues and questions concerning methodology to provide assurance, where warranted, that current laboratory results were relevant to natural environments.

Biotransformation tests are generally conducted to obtain information about the fate of chemicals in certain environments. For example, shake flask and biochemical oxygen demand type tests may simulate surface water environments, whereas acclimated sludge systems are used to simulate a waste treatment process. Deciding whether the information obtained is relevant to a natural environment often depends on whether some assumption is unknowingly violated or whether a particular problem exists with testing a given chemical in a certain manner. Fortunately, specific problems associated with the evaluation of insoluble or volatile chemicals can usually be surmounted by making small modifications in the method used to assess water-soluble chemicals. However, the procedural changes required to deal with different physical-chemical properties cannot be generalized and, as such, each must be dealt with on a case-by-case basis. This is particularly true when complex test systems are employed because small changes in procedures, such as adding specific types of nutrients or solvents, may affect the environmental relevance of the test.

It was agreed that chemicals which can be used as a sole carbon source by a naturally occurring microbial population will generally be nonpersistent in the environment. If acclimation, cometabolism, or specific environmental conditions are required to elicit biotransformation, extrapolation to natural conditions becomes more complex. The potential hazard posed by a particular chemical depends on its environmental release rate relative to its transformation rate, plus its toxicity. Thus, the results of a simple biotransformation test alone cannot be used to determine unacceptability, any more than rapid biotransformation can demonstrate acceptability: all elements of the hazard evaluation process have to be considered.

> "Biotransformation test methodology is developing rapidly and will continue to do so for the forseeable future. Thus, individual tests cannot be listed as standard or consensus methods for use in all cases."

The tier testing concept and consideration of environmental input modes were reflected in many of the test methodologies discussed. The testing schemes discussed generally proceeded from simple, static liquid cultures to multiple-component and dynamic test systems, as needed to reach defensible decisions. The tier approach generally assumes that compounds having intermediate degrees and rates of transformation in simple biotransformation tests will be subjected to additional testing. Conversely, microbially refractive compounds, and readily transformed compounds, will be detected early and appropriate decisions can be made with a minimum amount of valuable technical effort. Regardless of the test employed, the microbial assemblage (i.e., inoculum) used should fairly represent a conservative boundary of the habitat receiving the chemical.

The relative merits of simple versus complex tests were subjected to considerable discussion without complete resolution. It was pointed out that refinement of laboratory test methods will not provide guaranteed field predictability, because increasing the complexity of laboratory experiments does not necessarily improve the interpretability of the data obtained. However, when a simple test identifies a compound with persistent properties, ^{14}C studies in multiple-component systems may become appropriate. The development of more elaborate tests (i.e., microcosms) may also be useful for validating mathematical models that predict the fate of chemicals. Thus, tiered methodology should be applied as an iterative process wherein the sensitivity necessary would be determined by the margin of safety required in a hazard evaluation program.

> "Accurate prediction of the EEC is a critical aspect of a hazard evaluation for chemicals in aquatic environments. Quantification of biotransformation processes that alter the EEC is necessary to obtain reliable predictions."

It was recognized that nonquantitative biotransformation methods are useful for identifying possible metabolites and for early-tier "pass-fail" decisions. However, once even simple mathematical models are employed to forecast the EEC, the rates for all important transformations must be described by some mathematical formalism. Ideally, the algorithms used to describe biotransformations should be rate equations from which rate constants can be derived. In order to attain this goal, several questions dealing with biomass activity, substrate concentration, acclimation, bioavailability, and the influence of other environmental parameters must be answered before rate constants can be widely used to assist in predictions of EECs. Also, since rate constant data have been developed for only a few transformation pathways and sites, the reproducibility of the rate constant approach needs to be verified for a wide variety of compounds and environmental habitats. Undoubtedly, variations will be identified during this verification phase. These variations should be evaluated and the confidence intervals should be determined because this process will be useful in setting boundary conditions for the application of kinetics. But, for now, the science of forecasting EECs will be better served if attention is focused on the similarities observed between different chemicals and habitats, not on exceptions to the rule. Consequently, developmental efforts to quantify biotransformation processes should proceed on a broad front in concert with the other programs being developed to forecast EECs.

"Rate-limiting factors, other than the substrate concentration and the active biomass density, can alter biotransformation processes and must be evaluated to precisely extrapolate laboratory data to field conditions."

The influence of major environmental variables, such as oxygen gradients, pH, and temperature, on biological transformation rates is poorly defined at the present time. However, this shortcoming does not substantially affect the utility of biotransformation information for extrapolation purposes, because the accuracy of the estimate only needs to be within approximately an order of magnitude to aid initial estimates of the EEC. Hence, detailed studies on the influence of environmental variables are generally warranted only for research purposes and for sensitive cases where the safety margin is narrow. Most other situations may be resolved by assuming that "worst-case" conditions exist.

"The value of quantitative models to predict environmental exposure concentrations and input rate functions was strongly reaffirmed."

It was generally agreed that mathematical techniques are useful for making EEC determinations, but the models should be kept as simple as possible. The application of mathematical modeling to predict point source concentrations of pentachlorophenol illustrated the value of using individual chemical properties of a compound to predict its ultimate environmental distribution. By using a materials balance approach, the distribution of pentachlorophenol and its associated impurities was clearly identified. It was interesting that the mathematical model employed utilized the same principles used to design chemical manufacturing plants. Since it was demonstrated that this approach has considerable value in predicting point source contributions and because a large data base is available from the American Institute of Chemical Engineers, further studies using the approach should be undertaken in order to predict the environmental concentration of different chemicals.

Recognition that the models used to make estimates of environmental effects and environmental concentrations for input to a hazard evaluation process have not been field validated continues to be a source of concern. For example, extrapolation of most biotransformation data to predict ecosystem concentrations is only roughly quantitative at present. Nonetheless, because of the state of the art and the limited scientific and capital resources available, hazard assessment decisions must be based on these data. Thus, refining existing models and validating their environmental applicability is undoubtedly necessary. Yet, increased model refinement and more sensitive measures of chemical effects on organisms may not be generally warranted because the significance of any interaction must ultimately be judged by the nature of the effects measured in an ecosystem. Hence, chemical fate studies must always be balanced against the severity and reversibility of any potential effects associated with the introduction of chemicals into the environment.

> "The bioavailability of chemicals to higher aquatic organisms appears to be determined by the solute concentration, but the factors which control bioavailability to microorganisms have not been rigorously defined."

The question of bioavailability relative to the techniques discussed was addressed at length because the conceptual bases for hazard evaluation as developed by these workshops all relate to a common principle. When a chemical enters the environment, it is distributed into different compartments according to its physical and chemical properties. This distributional phenomenon means that only a percentage of the compound will be available for interaction with living organisms existing in a particular compartment. For the purposes of obtaining quantitative data for predictions of estimated environmental concentrations, we have implicitly assumed that bioavailability refers to that fraction of chemical which is soluble in water. This is because all equations for the derivation of the rate constants depend on accurate measurements of concentration. It is conceivable that other routes exist for the accumulation of hydrophobic molecules in living cells, such as phagocytosis, pinocytosis, and direct interaction with cell membranes; however, in the absence of quantitative data, speculation on the contributions of these routes of entry to the living world are premature. Consequently, bioavailability and all environmental processes are assumed to be a function of the "true" concentration of the chemical in water.

FUTURE RESEARCH NEEDS

The success of these workshops has partially resulted from consideration of specific topics most relevant to existing scientific and regulatory needs. The first two workshops focused on general approaches for evaluating chemical contaminants in aquatic systems with emphasis on aquatic toxicology and water quality criteria. This workshop focused on a major type of reaction (biotransformation) that often is critically important to selected environmental hazard assessment programs. There are many other aspects of the hazard assessment process that also need future attention similar to that given biotransformation.

Several participants at this workshop stated that a highest priority for research needs should be given to methods for assessing the mode, amount, and rate of input of chemicals to the environment. Prominence was given to this area because it is key to the development of reliable estimates of expected environmental concentration. It is from the point of first entry that the hazard assessment process begins. Hence, the physical processes of advection (i.e., transport) and dispersion (i.e., mixing) in aquatic systems would also need to be considered. Methods for both homogeneous (true solution—dissolved) and heterogeneous (solution and solids) transport and mixing need to be developed and refined. Particular emphasis should be given to those areas of hydraulics and hydrodynamics that relate to developing concentration-time profiles for chemicals that remain in solution or are associated with natural water particulate matter, or assume both forms. Combining the procedures used to estimate initial environmental concentrations with advection-dispersion of solids and solution behavior has merit because it should define the boundary conditions which govern the "worst-case" situation that arises immediately after entry of a chemical into aquatic ecosystems. At this time, the chemical can be regarded as being conservative, since the only reactions that have had an opportunity to occur are basically the association of the chemical with suspended solids and sediments. Thus, all aspects surrounding both intermittent and continuous inputs of chemicals to aquatic environments will require carefully integrated research attention.

Another topic that will demand future research effort is review and verification of existing models. It was generally agreed that environmental chemistry-fate models should be as simple as possible. That is, they should employ the least number of variable coefficients needed to adequately describe the environmental behavior of a chemical, yet retain a sufficient degree of confidence to permit reliable decisions to be made. Thus, the focal point of future research should be the merits and problems inherent with each type of modeling approach relative to their utility or applicability to environmental hazard assessments.

Overall model validation should be discussed in terms of the reliability of the information needed to make an environmental hazard assessment. That is, the sensitivity of the decision-making process should consider the reliability of the data being employed. For example, the decision maker might arbitrarily increase the expected environmental concentration by a factor of 10 and then consider whether errors in the data provided would change any decisions. Ultimately, attempts might be made to quantify the confidence intervals about the EEC and no-observed-effect concentration at any level in the decision-making process, particularly when the safety margin is narrow. In any case, the sensitivity of the decision must consider both the consequences and "reversibility" of an erroneous decision. Thus, some aspects of risk analysis will eventually need to be incorporated into methods of hazard evaluation.

The practical aspects of environmental hazard assessment also need more attention in the future. For example, the development of water quality standards by individual states based on National Water Quality Criteria immediately tests many of the general assumptions made concerning real-world extrapolation of data. Such an approach would be applicable not only to the United States, but also to other countries where diverse opinions exist regarding allocations of input loads of contaminants into particular water bodies (e.g., the Rhine River). Conversely, methods to assess the recovery of aquatic systems after intentional reductions in chemical contaminant input rates need to be refined. If emphasis were placed on developing load-response relationships, it might be possible to predict the response of given aquatic systems to certain levels of chemical load reduction. More reliable and quantitative estimates of both potential damage and benefits might be derived if a series of such case studies was developed. Thus, numerous theoretical issues and practical problems in hazard evaluation can be identified that still remain as active research topics for the future.

LITERATURE CITED

1. **Cairns, J., K. L. Dickson, and A. W. Maki (ed.).** 1978. Estimating the hazard of chemical substances to aquatic life. ASTM Special Technical Publ. 657. American Society for Testing and Materials, Philadelphia.
2. **Dickson, K. L., A. W. Maki, and J. Cairns.** 1979. Analyzing the hazard evaluation process. American Fisheries Society, Bethesda, Md.

Author Index

Subject Index